Michael Macray

Adventures in Disaster

Adventures in Disaster

2nd Edition

Print Paperback ISBN-13: 979-8-218-27673-7

Print Hardback ISBN-13: 979-8-218-37621-5

Ebook ISBN-13: 979-8-218-48221-3

Dedication:
To Sage, Mac and their parents, Tyler and Kelsie.

Acknowledgments:
Among those who have helped me bring this book to fruition, are two in particular, to both of whom I am especially grateful.

To my Editor, Libby McMillan Henson, accomplished Author, Writer, Content and Idea Producer who managed to bridge the Atlantic chasm between my English and the American version of the same language, which, believe me, was no small feat.

And no less thanks to my dear friend Shawn Coughlin, without whose talents in the Graphic Arts and dexterity in the digital world, I would have been dueling the Digital Dragon (guarding the Publisher's Gate) with my slide-rule for Eternity. . . along with Sisyphus.

Table of Contents

Introduction

I am a poor man with a rich life history, for which I claim no credit.

Among the many who have inspired me – and they are legion – are Peter and Maria Matthiessen. Peter autographed my copy of his National Book Award winner, 'Shadow Country,' with an ancient proverb: "He who has no destination is never lost."

It was the perfect answer to my dear departed father's eternal question: "When are you going to get a proper job?"

I grew up in England during the Cold War following World War II, when the victorious Allies, united by a common foe, had instantly become enemies without one. Winston Churchill defined the transition in a speech delivered in Fulton, Missouri on March 5th, 1946, with the immortal words, "From Stettin in the Baltic, to Trieste in the Adriatic, an 'Iron Curtain' has descended across Europe."

For the next half century, the world was divided by the "Iron Curtain."

At the time, few understood the implication of Churchill's historic observation. The Soviet detonation of its own atomic bomb in 1949 stunned the Western world, which had assumed that the enormous resources invested in The Manhattan Project were way beyond the means of a nation like Russia, emerging from feudalism, a Marxist revolution and ravaged by four years of a brutally devastating war.

Little did we know, when President Harry Truman announced to an unimpressed Joseph Stalin at the Potsdam conference the successful detonation of the atomic bomb a few hours earlier, that Stalin already had working drawings of the bomb from his embedded spies: Klause Fuchs and the Rosenbergs, among others.

The detonation of the Soviet bomb just four years after Nagasaki leveled, in a sense, the playing field; at least to the extent that Mutually Assured Destruction (MAD) eliminated its use, triggering the Korean War and every subsequent periferal conflict under the Nuclear Umbrella, at least until the suicide bombers of 9/11 came into play.

So warfare became a peripheral skirmishing game, both sides using surrogate rebel organizations of the so-called non-aligned nations, all the while carefully disguised with 'plausible deniability' as cover.

On our side, the game was known as 'Poking the Bear.' On their side, these were piously labeled 'Wars of Liberation.'

As the Colonial era ended, a host of newly independent nations became targets of the Cold War adversaries, each side seeking to lure them into their ideological 'camp,' so to speak.

Our slogan was Freedom, theirs was Liberation, so you can understand how confused were the barely literate, hatchling nations. The 'lure' was aid, mostly military, which provided a market for old weapons, as the Cold War arms race propelled the Military Industrial Complex to produce new improved ones for the rich nations and superpowers.

Among the most strategic geopolitical targets was the 'Horn of Africa', the African continent's northeastern protrusion, which holds Somalia, and Ethiopia. Squeezed between these two countries is the minuscule former French Colony of Djibouti, home of the then-formidable French Foreign Legion, and a French naval base guarding the southern approaches to the Red Sea and Europe's energy jugular, the Anglo-French Suez Canal.

THE HORN OF AFRICA

There was a little room left over in Djibouti for a 'downtown' area where locals and commercial activities could feed from the military trough.

Across the Red Sea's narrow throat, known in Arabic as the 'Ba'ab El Mandeb' – the Gate of Affliction – lies Yemen, the southeast corner of the Arabian peninsula, and home to the Port of Aden, then a substantial British military base which guarded the northern approaches to the Red Sea and the Suez Canal.

As the West's post-war economies boomed, so did its appetite for 'black gold,' the vast oil reserves of the Middle East. . . and so did the geo-political competition for the West's choke point, the Suez Canal.

As monarchy after monarchy of the post-colonial domesticated regimes toppled to Soviet-inspired military coups, the West eventually woke up to reality and started to push back. At first unsuccessfully, by backing the top-down autocratic anti-communist regimes – like Iran, Cuba and Guatemala – and later, when that didn't work, adopting the bottom-up support of counter-insurgency: Nicaragua, Ethiopia and Afghanistan, among lesser-known ones, come to mind.

Like geology's tectonic ring of fire, the Cold War ring of fire defined the skirmishing hot spots between the competing East-West ideologies, across the Iron Curtain, so to speak.

Among the first to fall was Indochina, the former French Empire in the Far East, seized from the vanquished French by Imperial Japan in WWII, and reluctant to reinstall its previous Colonial masters. The battle of Dien Bien Phu (1954) ended French hopes at recolonisation, and opened the way for Uncle Sam to spend a couple of decades trying, unsuccessfully, to put Humpty Dumpty together again.

But I digress.

As civil wars proliferated, so did the humanitarian tragedies of famine, destruction, displacement, and lawlessness.
Enduring ethnic enmities, exploited by external influences and fueled by internal corruption all combined to disrupt the fragile evolution of nationhood in the hatchling nations, staggering like newborn calves in their post-colonial infancy.

The affluent West responded with "Humanitarian Relief," spawning a multitude of International, National and Non-Government Organizations (NGOs), funded by everything from begging bowls, church donations and rock concerts all the way to national and international taxpayer-funded organizations.

And so Humanitarian Relief, being morally unchallengeable, gained access to the heart of the conflict zones, and, of course, firsthand ground-level information...the feedstock of the national intelligence agencies, those bloodhounds of the information trail, who, needless to say, were never far behind.

As the relief business grew, so did the need for funding. Inevitably the search for funding led to the deepest pockets: governments . . . and the richest (and least regulated) sources were the intelligence agencies, to which, inevitably, fund seekers gravitated, and – not surprisingly – the Intel folk sought to infiltrate.

Besides the CIA, almost every US government agency has its own intelligence outfit.

One of our paymasters was the Office of Foreign Disaster Assistance – OFDA, a subsidiary of State, which had its own piggy bank – so didn't need an act of Congress to fund an operation; essential when responding to emergencies, and convenient for more subtle operations.

The Horn of Africa, among others, was a strategic hot spot. Both as a portal to sub Saharan Africa and a gatekeeper to the Red Sea and the Suez Canal.

The 1973 Yom Kippur War reshuffled Middle East allegiances. Egypt, a Soviet surrogate, made peace with Israel and, lured with substantial benefits – among them Nobel Peace prizes – came over to our side, whereupon a Soviet-inspired military coup toppled and murdered our old ally Emperor Haillie Selassie, and Ethiopia moved into the Soviet camp.

Sudan, with its long border with Ethiopia, came into play as a handy base from which to support rebellions in Ethiopia's northern provinces of Tigray and Eritrea, and about which the Sudanese professed no knowledge. Coincidentally, rebellion in southern Sudan had bases conveniently located across the border in southern Ethiopia, about which the Ethiopian regime was officially ignorant.

It was a sort of 'round robin' game of 'you kick my butt and I'll kick yours!'

All of which helps explain why one day we were smuggling nuns and money to Ethiopian Rebels besieging garrison towns like Shiraro and Asmara, and the next day flying emergency supplies into besieged garrison towns like Juba in southern Sudan.

But enough of the preamble, time to get on with the story... and Aden, South Yemen is a good place to start. Enjoy!

- Mike

CHAPTER 1:
Aden, South Yemen, and Abu Dhabi

ALI SUBEHI (SOO-BAY-HE)

Aden, 1966

Harry was my boss. He was the Assistant Planning Engineer at the BP refinery in Bureika, Little Aden, about 20 miles across the bay from the busy Port Town of Aden.

Harry was an ex-SAS guy who in WW2 had trained to be parachuted at night into the mountains of Norway. The mission was to sabotage the Heavy Water plant that the Allies feared was part of the Nazis' program to develop the Bomb. The exit strategy was slim to none, the kind of challenge that appealed to SAS types.

I remember him describing a training exercise on a dark moonless night in Scotland's Cairngorm mountains, standing with his squad in full battle rig waiting for the signal to drop through the open hatch in the floor, as the plane throttled back, reducing speed for the drop. The aircraft had suddenly lurched violently and started to climb with full power. Just behind the open hatch was a big pile of snow. The plane had bounced off the top of a mountain.

Harry was an erratic, high-energy chap who was always on the go, and always entertaining. He spoke good colloquial Arabic, which included dramatic hand waving, essential to the local language. He was also an expert on the nomadic tribes that roamed the Subehi, the supposedly uninhabited desert between the coast and the highlands of the Hadramaut, in which he had identified some 700 tribal groups ranging in size from half a dozen up to a hundred or more. Each tribe had a distinct identity and its own tribal leader, or Sheikh.

At the refinery, after checking in briefly to the air-conditioned office, Harry would 'go onsite,' which involved wandering around the property, chatting with the coolies, contract laborers shipped in daily for the menial, unskilled work which earned them around 50 cents a day. Most were Bedouin, who could live on 5 cents a day, and often were unfamiliar with technology as sophisticated as a door handle. I kid you not!

I remember one occasion when a very nervous new sweeper was creeping around the office with his broom when Harry's phone rang. Startled by the ring tone, he stared bewildered at the phone. Never one to miss an opportunity, Harry picked up. "It's for you," he said, holding the handset out to the bewildered sweeper, and coaxing him to hold it up to his ear. On hearing the voice, the coolie freaked out, beating the phone with his broom amid a volley of Arabic curses, believing that the voice was an evil genie hiding in the phone!

Harry was well-known and revered among the locals as Ali Subehi – thereby granting him membership in all the nomadic tribes of the Subehi. Out and about in the refinery, he would home in on a likely coolie, and engage him in conversation, enquiring about his tribe, and his Sheik, to whom Harry would convey his greetings.

Harry would also inquire about the number of their male offspring, the number of their camels, and, of course, their marital status: new wives and consorts essential to the future of the tribe, as any student of the Abrahamic covenant will understand, (via Ishmael, not Isaac, of course).

On several long weekends, Harry took me deep into the desert in his homemade dune buggy. We would hang out with some tribe or other, adapting seamlessly into a lifestyle that had not changed since the time of the Prophet.

It was a time warp.

Apart from the odd jacket for status, usually with an empty pack of

State Express 555 cigarettes protruding from the breast pocket for show, there was nothing to show that in a few hours Harry and I had not traveled from the 20th Century back to the Middle Ages.

To the nomadic Bedouin, there were 'Wallies' everywhere. A Wally is a holy man, a saint or hermit, some living and others entombed but still requiring an offering or blessing to ensure safe passage through their domain. There were also, to the bedou, abundant genies or spirits, usually just mischievous but sometimes evil, all of whom had to be acknowledged and appeased with an offering or an incantation.

When bedding down in the desert sand for the night with our Bedouin hosts, we followed protocol by surrounding our sleeping mats with a ring of stones to keep out scorpions and evil spirits . . . in all seriousness.

Among his other eccentricities, Harry had a habit of disappearing for a few days at a time, which somehow no one seemed to notice, until one day he just disappeared altogether. After a couple of weeks, the BBC World News reported that a British spy had been arrested and was jailed in Sana, the capital of North Yemen, a separate country at the time, and occupied by Egypt's Gamal Abdul Nasser, a Soviet surrogate.

Needless to say, it was, of course, Harry.

With his local knowledge and connections, he had managed to chat up one of his fellow inmates or perhaps one of his jailers, and get the word out. He was diplomatically extracted from his jailers, but alas, with his cover blown, BP had to fire him.

Harry had been an agent of British intelligence, and was tracking the supply lines of weapons being infiltrated through the 'uninhabited' Subehi desert, to sustain the revolution in Aden. This was my first introduction to the dual world of Appearance and Reality. Later, I got to know several SAS chaps on 'leave of absence' from "The Regiment." Of course this meant that Her Majesty's Government could 'plausibly deny' responsibility for any mischief of their doing which might come to light.

They were reputedly well paid by the Saudis in any currency they chose, including gold. But the real advantage of knowing these guys was the stuff to which they had access. As the security situation degenerated, with the impending and well-publicized British withdrawal, we were advised to "carry," which most of us did. The permit allowed 200 rounds a year: not much use if one wanted to practice.

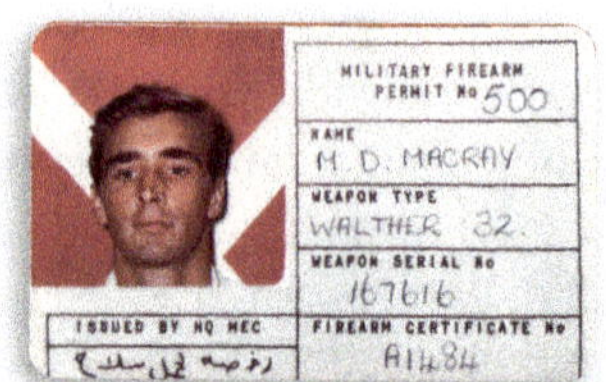

Knowing chaps in the SAS helped a lot. They would show up with boxes of 1000 rounds, for a picnic in our boat. Target practice was on the beach or the rocks, shooting crabs lured by leftover scraps from the picnic . . . but with very accomplished instructors.

As the withdrawal progressed, remaining expatriates became targets. At one point, rebel leaflets announced the assassination of "one expat a day" until independence. This was a bit nerve-wracking because for several days they did bump off an expat a day, and there were more days left than remaining expats.

Meanwhile, North Yemen was in a civil war between the military – bolstered by a sizable Egyptian contingent which garrisoned the towns – and the Bedouin, loyal to the deposed Imam who controlled the countryside, and especially the hilly 'upcountry.'

The specialty of the SAS is operating behind enemy lines. So the chaps would periodically disappear for a week or two before reappearing at one of their habitual waterholes, like the bar at the Crescent Hotel, a gathering place for the International Press Corp. Once settled behind their favorite beverage, they would express thinly disguised astonishment at reports trickling in from the field of a major ambush of an Egyptian column by loyalists in Wadi 'Rum Dum or such' . . . interestingly, half a century later, the Saudis are still actively involved with events across their southern border in Yemen.

Plus ça change, plus c'est la même chose!

Map of South Yemen

The Subehi, is the desert area between the coast and the highlands to the north.

Piper's Pies & Other Secret Weapons

Aden. Circa 1966

The BP Oil refinery in Little Aden, Yemen – about 20 miles across the bay from the Port of Aden – was a popular target for pyromaniacs of the Revolutionary kind.

Refineries make spectacular targets.

It takes a very small amount of strategically-placed explosive to make a really big impact and get everyone's attention, especially the bigwigs of the company. The favorite time for an attack was late on a Saturday evening, because local Arab workers got double pay cleaning up the mess on Sunday. There was also the added advantage of annoying the expatriate staff who didn't get overtime, and who were usually having a good time at the Club, all dressed up in their party clothes.

The perpetrators of these attacks were not dignified, with powerful labels like 'terrorists' . . . rather they were called 'dissidents' and were the subject of ridicule. Most of the attacks were on pipelines and other relatively 'soft' targets outside the refinery fence, and were thus fairly easy to isolate and get under control.

In addition, Shell Oil Company had some installations on the foreshore, and it was not unusual following the sound of a late evening explosion for someone to shout, "It's OK, it's the Shell terminal," and everyone would breathe a sigh of relief, and go back to the bar and whatever they were drinking.

Nevertheless, they – the dissidents – did succeed in keeping us on our toes and not a little nervous.

Along with these periodic attacks was a systematic civil disobedience program that called for endless strikes. There were two main revolutionary entities vying with each other to demonstrate who had the most support of the people.
They were the National Liberation Front, or NLF, and the Front for the Liberation of Occupied South Yemen, or FLOSY as it was known.

The former was based in the mountainous regions of Muscat and Oman, sponsored by Mao Tse-tung's China, and made up mostly of Bedou, as the nomadic tribesmen are known, and whose talents reside in the use of the sword and the gun. FLOSY, on the other hand, was backed by the USSR via their Egyptian surrogate Gamal Abdul Nasser, and drew its support from the 'trousered,' or urban Arabs of the commercial port of Aden: who drove cars, used telephones and did business.

I will leave it to my reader to guess who won the brief civil war. Suffice it to say, though 'the pen may be mightier than the sword,' in South Yemen, the phone was no match for the AK47.

The NLF would call for a three-day strike to demonstrate the 'people's solidarity' with the NLF, after which FLOSY would call for a four-day strike to demonstrate their solidarity with the people. Needless to say, no local workers came to work when a strike was called by either party.

The consequences for a local breaking the strike – regardless of who called it – could be hazardous to one's health or even one's life. An extreme version of Cancel Culture if you will.
This meant that sixty or so expats had to keep the refinery running by working twelve-hour shifts while the thousand or so local employees were on strike.

For example, the Crude Distillation Unit – to which I was assigned – would have two guys on 12-hour shifts, where normally there would be three eight-man crews working eight-hour shifts. Being an engineer, my job was maintenance. This involved running around the plant with an oil can, squirting the bearings on pumps that made inappropriate noises, or hosing them down with cold water if they looked overheated. Meanwhile, my cohort Ted – a chemist by trade from the Process Department – worked in the control room, running up and down the control panels, tweaking knobs to keep the needles in the middle of their dials . . . both of us praying fervently that if something had to go wrong, it would be on the next shift.

Most of the time we got away with it . . . except for once when an explosion in the control room got everyone's attention, and how!

All the alarms went off, red lights flashed everywhere, sirens wailed, hooters howled, fire alarms jangled, and in no time the ever-ready Refinery fire brigade arrived. The bomb squad was alerted and arrived soon after, from the nearby British military base in Falaise. And of course, all the bigwigs showed up to see what happened.

Picking through the debris of broken glass and twisted metal with forensic thoroughness, the officer in charge of the bomb squad honed in on some soft brown material about the consistency of thick gravy that had been distributed fairly liberally about the control room. There wasn't very much of it in any one place, but it was all over the shop.

I remember him examining a sample closely, smelling it, poking his finger in it and then finally tasting it. He paused, puzzled, sniffed and tasted a second sample and then with an air of diagnostic triumph announced:
"It's Steak and Kidney pie!"
No kidding.

There are no meal breaks for the shift workers, so the control room was furnished with a fridge and mini-oven for the convenience of the shift workers. 'Someone' had put a Piper's brand can of steak and kidney pie in the oven, omitted to puncture it and – distracted by other pressing priorities – forgot it . . . until it exploded.

For those unfortunates who neglected the study of thermodynamics in school, it is the awesome power that Mother Nature releases when heat turns water into steam. It can trigger an industrial revolution, or make a bomb out of a can of steak and kidney pie.

The most popular radio program on the Armed Forces Radio in Aden was a show by a young officer from 45 Royal Marine Commando, who had a talent for writing topical parodies of well-known Irish ballads, which he sang to the accompaniment of his guitar. The main character in his ballads was a 'dissident' named Mohamed Abdoo, who was always boasting of his latest attempt at blowing up this or that, and then demonstrating his other talent: dancing.

The story is immortalized in the 'Ballad of Mohammed Abdoo The Great Refinery Raid.'

The refrain as I recall went something like this:
"For I'm a handsome golly
and dissidence is my folly,
With a hand grenade and a bunch of qat*
I'll tell you what I'll do . . .
I'll chew the qat and blow them flat
and head for the disco too.

And there it is that I'll drop 'em dead
as I waggle my dancing shoe . . .
But nobody knows that my secret lies
in a can of Piper's Pies."

*For those unfamiliar with the popular recreational narcotic in Yemen and the Horn of Africa where Islam prohibits alcohol, Qat is the bitter-tasting leaf of a scruffy bush indigenous to the region, and a powerful stimulant: an 'upper' like cocaine. Sold in bunches, mostly on payday, it is chewed but not swallowed, like tobacco in Mississippi, coca leaf in South America or betel nut in Asia. Qat contains the alkaloid cathinone, a stimulant which induces euphoria, erratic behavior, fast talk, bug eyes, a wad in the cheek (like a hamster), and black teeth.

The Great Refinery Raid

Aden, South Yemen, 1966

Oil refineries, it goes without saying, are a fire hazard second to none; up there with munitions and fireworks factories. The safety precautions were draconian; possession of matches or a cigarette lighter inside the fence was grounds for instant dismissal; and rubber-soled footwear was mandatory, lest a metal nail or toe cap kicking a stone should cause a spark.

The obsession with fire prevention was nowhere better characterized than by a line item in the company store's inventory, which read: 'Fully automated non-electrostatic grass cutters: . . . . 30.'

They were goats.

The herd roamed free inside the refinery fence, eking out a meager existence by eating any plant life that survived the harsh desert environment long enough to dry into a fire hazard. The fire department moved water troughs around to places where the goats were needed.

With over a million tons of crude and refined petroleum typically stored at any one time in the tank farm, it was inevitable that the pyromaniacs of the Revolutionary Front(s) would find the BP Refinery an irresistible target.

And indeed they did.

Most of the attacks took place on Saturday nights, when the senior (mostly European) staff were partying at the Club. This annoyed the expatriates, spoiling their Saturday night revelry, but was popular with the local workers who got double pay for the weekend cleanup.

The sabotage started on the pipelines that ran a mile or so from the refinery to the foreshore, where the tankers discharged crude oil and took on refined products. This was not a big deal as the damage was quickly isolated and repaired.

Not so on the night of the great refinery raid.

The fireball came from the heart of the tank farm, where close to a million tons of highly combustible stuff was stored. To understand the nature of the situation, it's necessary to get a bit technical so bear with me.

The tank farm covered well over half the refinery land area and was built like a giant breakfast waffle with the storage tanks set down in the square depressions (called bunds). On top of the elevated earth dikes separating the bunds were the service and access roads, some 8 to 10 feet above the bund floor where the tanks sat. This was designed to contain the oil in the event of a tank failure.

On the night of the attack that's exactly what it did.
Around 9 pm, there was the familiar boom of an explosion, but to the horror of all who heard it, a great fireball erupted from the middle of the tank farm. The charge had been well placed, blowing out a control valve and nozzle at the base of one of two 52-foot-high, five thousand ton capacity tanks containing JP4, a high grade military jet fuel. That's just shy of one and a half million gallons capacity for each tank. The fuel gushed out, flooding the bund and spreading the fire, which soon engulfed the second tank, licking hungrily at its wall.

The hooter went off, its mournful wail drowning the puny desert sounds, and it was all hands on deck.

52 Foot Tall Tank Blowing Its Top

In various stages of inebriation, we raced to the Refinery, in most cases shedding our party attire for the standard white cotton work boilersuits. The fire department was already on the scene when we arrived at the inferno.

Water is pretty useless for oil fires because the oil floats and continues to burn, so the trick is to blanket the oil with foam, cutting it off from its lifeblood: oxygen in the air.
By the time we arrived, the Refinery Fire Department were doing their thing, spraying water into the flames and pouring foam onto the burning kerosene, a lake that now engulfed the wounded tank and its

twin with rings of fire. The heat was blistering. It was impossible to stand with a fire hose for more than a minute or two on top of the bund without roasting, or worse, bursting into flames, so we hosed each other down to soak our cotton boiler suits, and lay down on the outside bank of the bund, like soldiers in a fire fight: pun intended!

It takes two guys to control a high pressure fire hose, which is like holding a rocket with 40 or 50 pounds of thrust; if it gets away from you, it thrashes around like a giant wounded snake and the heavy brass nozzle can break arms and legs.

Water is sprayed into the heart of the fire in a mist which evaporates into steam, sucking with it a lot of the heat, and, in theory, providing a blanket of steam to reduce the amount of oxygen available. The steam blanket theory works well in enclosed spaces, but not so well outside, where the steam gets carried up and away with the flames . . . but it sure does suck the heat out of the heart of the fire.

Along the inside of the bund were racks of 4, 6 and 8-inch heavy-duty steel pipes. These, of course, had been isolated, but were still full of fuel, which was cooking quietly in the flames. Every so often, one of the pipes would burst, causing a massive explosion and fireball, sending everyone tumbling down the bank clutching their hoses and whatever else they valued.

Strange to say, there was something of a festive air, despite the fact that we were surrounded by combustibles, not least four tanks of high-octane gasoline in the next bund just a few yards away. All the big shots were there, urging us on with encouraging remarks like: "anyone got any chestnuts?" which seemed funny at the time precisely for its absurdity.

As the night wore on, we slowly gained on the inferno, the blanket of foam creeping inexorably over the flaming kerosene and gradually reducing its burning area. Every so often, the volatile vapors that managed to find a way through the foam drifted into the fire, triggering a flash-back: a great 'Woompf' accompanied by a fireball which sent us all tumbling back down the bank.

By morning, the fire was out on the lake, which was now covered with a light brown blanket of froth, but fire was still burning inside one of the tanks. This was a problem. The only way to access the fire was to climb the gantry scaffold between the tanks and pour foam over the 52 foot high tank wall down onto the flames. That meant wading through a foam-capped lake of hot jet fuel, dragging fire hoses.

THE GANTRY BETWEEN THE TWIN 52FT. TALL TANKS.

Understandably, nobody wanted to do it, especially the local Yemeni firemen who were mostly quite short.

Still running high on adrenaline and low on common sense, Electrical Pete (the electrical engineer) and I volunteered, and waded into the lake, dragging a fire hose and a couple of reluctant members of the Refinery Fire Brigade with us. The kerosene was pretty warm, and chest deep, but we made it across and climbed up the gantry between the tanks and peered over the steel wall into the well.

Most of the oil had poured out into the bund so the floating roof was well down the tank, but the fire was licking hungrily around the perimeter, having devoured the seal to get at the 'juice'!

Explosions occur when air and vapor mix, so oil storage tanks are fitted with shallow conical roofs that float up and down on the oil to eliminate the air chamber.

It looked like a fairly simple process to pour the foam down the side wall of the tank and let it do its thing as it worked its way around the perimeter snuffing out the flames. But oh no . . . the fire had another agenda.

The combustible vapor accumulated over the foam. Contained by the walls of the tank it built up, creeping around the perimeter of the floating roof until it found the remaining flame . . and then exploded, blasting a fire ball up into the sky. Each blast threw us back from the tank rim, the flames missing us by a mere foot or two.
Fortunately, in the adrenaline-soaked heat of the moment, it never occurred to us just how easily we could have caught fire in our kerosene soaked cotton boiler suits....with no escape except to dive into the hot kerosene lake fifty feet below!

It took a couple of hours and several flashbacks before we finally got the fire suppressed. But after only an hour or so, the warm oily sensation of the kerosene began to change into an itch.
The itch grew stronger and stronger until it started to burn. I looked over at Pete and noticed that he too was hopping around and scratching like a monkey in a flea market. By the time we were done, we were both on fire.

Worst of all, the affliction was concentrated on those most sensitive areas where the skin is thinnest: the armpits and yes, those other places that it would be indecorous to name. Racing down the gantry, dragging the fire hose along with the Adenese firemen, we plunged into the foam-capped kerosene lake.

The level had risen during the time we were up the gantry.
Not a problem for six-footers like Pete and I, but a couple of the local lads we took with us were now chin-deep and one of them stumbled and went under.

Kerosene is much lighter than water, so impossible to swim in. Fortunately we were dragging the firehose, which turned out to be the lifeline for our diminutive fellow fireman. We were able to drag him coughing and spluttering to the bank and pack him off to the Refinery Hospital.

Having discharged our obligations, or so we thought, we attended to our own problem, the fires in the groin, and set off at full gallop to the nearest showers. Pete jumped the gun and took off just ahead of me, shedding his garments like Hippomenes tossing the golden apples of Hesperides in Atalanta's race* with me close behind.

When no amount of soap and water cured the condition, though it did quench the fire so to speak, it was off to the company clinic, never mind the embarrassment of where the problem was. I learned soon enough of a condition called Kerosene dermatitis. Kerosene and other powerful solvents extract the natural oils of the skin exposing the nerves with painful consequences. Happily, soothing balms eased the pain until Mother Nature could restore the oils.

Sadly our diminutive fireman friend did not survive. Unlike water, kerosene inhalation is fatal.
In retrospect, I cannot help but reflect on just how fine is the line between the here and the hereafter.

On this occasion, it was about 9 inches, the distance between 6 ft. 3 and 5 ft. 6.

Crumpled steel tank wall, lamp post and pipes all buckled from the heat of the inferno.

TWO DAYS LATER BOATING ON THE JET FUEL LAKE.

Buckled steel pipes from the heat of the fire

Footnote:

*In Greek mythology the infant Atalanta, abandoned by her father, was left to die in the woods. Suckled by a she-bear at the behest of a goddess, she survived, and endowed with great fleetness of foot, became a great hunter. Reunited with her father who was anxious to marry her off, (presumably for a good dowry), she agreed to marry any man who could outrun her. Many suitors tried and failed, and were killed as the downside of the deal: The King didn't like losers. Then along came Hippomenes, who had the 'hots' for Atalanta. Smarter than the average Greek, he first consulted with Aphrodite the goddess of love, who gave him three Golden Apples of Hesperides (oranges?) and a race plan. Each time Atalanta passed him he dropped an Apple. Unable to resist, she stopped to pick up the golden apple, allowing him to take the lead. Carefully managing his assets, Hippomenes was able to pip her at the post with a timely toss of the third Apple... and win the elusive prize. What is it about temptation, apples and women...? Oops!

The Party went off with a Bang

Aden, 1967

The Goldmohr Beach Club was a scruffy affair at the end of a road which skirted the outer base of Crater mountain, along the beach from Steamer Point. Inside the caldera of the extinct volcano – known, not surprisingly, as the Crater district – lived the bulk of the town of Aden's Arab population.

Densely populated, the area was a warren of narrow streets, shops, souks, repair shops and restaurants where the men gathered to chew the immensely popular and ubiquitous narcotic leaf 'qat.' There were only two ways in or out, one over the crater wall from Ma'alla on the Harbour side, and the other along the Khormaksar road, where eons ago the wall had collapsed, spilling its boiling brimstone into the Indian Ocean.

Crater was off limits to expats, since about half of the adventurous (or foolhardy) souls – depending on your perspective – who ventured in seeking bargains in the souk had come out dead or not at all.

But Goldmohr was pretty secure. Two or three checkpoints and a two-thousand-foot red rock wall separated the assassins from the sunburned white guys. It was here at the Goldmohr Beach Club that I met Andrew, an official in the British administration of the Aden Protectorate. A decade or so my senior, he had started his career as a 'Colonial Administrator' in Nigeria, and coincidentally my father was his first boss. Thus was a relationship born of a mutual coincidence.

Port of Aden
Crater district and Indian Ocean

He was older and married. I was young, single, and, it seems, amusing, so a friendship developed, which led to a party invitation at Andrew's apartment in Ma'alla. At the time, a kind of martial law prevailed, which banned gatherings of more than twelve expats without official sanction, since such congregations were deemed targets of the dissidents (they had not yet been dignified with the elevated status of 'terrorists'). And so it was that I arrived at my new friend's flat for the party all dressed up in my new tailor-made tropical suit by Chan Tuck, the Chinese tailor.

The flat was a modest one bedroom on the third floor, with a shoebox-shaped living room overlooking the so called Ma'alla Straight, the main drag between rows of mid-rise apartments to Steamer Point, the shopping center for the port of Aden.

In its halcyon days – prior to the June '67 six-day war – all the shipping traffic (passenger and freight) coming and going through the Suez

Canal bunkered in Aden to take on fuel and disgorge waves of shoppers for duty- free cameras, stereos and booze, which, back home, carried punitive 'luxury' taxes.

The short side of the 'shoebox' apartment led onto a small balcony overlooking the Ma'alla Straight. Along the side walls was parked the room's furniture, chairs, sofa and a bookshelf, leaving the center of the room open for guests to mingle.

On both walls were a series of triple-branch candelabra with those little twisty light bulbs made to look like candles. These, together with a few strategically placed lamps, lit the room. In one corner of the back wall was the entrance door, and in the other corner was a temporary bar set up for the occasion.

A couple of hours into the party there was a massive explosion.

It wasn't exactly loud, more like a huge thump from all sides by giant pillows followed by stunned silence. A cloud of black smoke curled and dissipated through the room; everything froze, including time, which stopped.

I remember a loud ringing, more in my head than in my ears, and then slowly, very slowly, sounds, cries and voices distorted and crackling like a broken radio, penetrated the ringing in my ears. Such moments – when time stops – last long; until slowly the wheels of time start turning again, in a kind of bizarre slow motion, gradually picking up speed, often to hyper levels, as adrenaline kicks in.

The room was dark: lit only by a single bulb on one of the candelabra dangling from the wall by its wires. It was the sole survivor of the dozen or more light bulbs burning before the blast. Shredded books, splintered furniture and bodies littered the floor of the apartment.

I remember a young woman slumped in a heap, her skirt blown away, staring in stunned disbelief at a wooden chair leg which impaled her thigh. This stake protruded exactly six inches on either side of her thigh. My first reaction was to pull it out, which thankfully I did not attempt. Instead, being, it seemed, the only one left standing, I ran out to get help.

The explosion had not gone unnoticed, and by the time I got down to the street, there were a couple of patrol vehicles already closing in on the building. The blast had blown out the windows so the site of the explosion was no secret. I led them up the staircase to the apartment.

I remember the rage that overwhelmed me. Had I had a gun, I would not have hesitated to pull the trigger on any local who showed a glimmer of satisfaction at what had just happened.
I learned this about myself in the crucible of fire, and ever since have been reluctant to judge those who over-react in such circumstances. I hold in utmost contempt those who have never been there, and yet presume to judge those who have.

It had not yet occurred to me how lucky I was to be alive. My white shirt was torn and bloodied, but I was only aware of superficial scratches from flying debris. It was not 'til the next day that I discovered, to my embarrassment, that I had a piece of Czech steel in my right buttock . . . surgically removed, eventually, by the venerable company doctor, and preserved for some years in a small glass bottle.

I learned that of the nineteen people in the room, thirteen went out on stretchers, half of them dead or permanently disabled. Only six of us walked out of that room.

The device was a Czechoslovakian jumping jack, an anti-personnel fragmentation bomb designed for confined spaces, with the specific intent of inflicting maximum damage to people. It had been placed in the bookshelf, behind a bunch of paperbacks, spring-loaded and timed to 'jump' out into the middle of the room and explode.

Later we learned it had not worked as intended.

The bookcase, which was two thirds of the way down the side wall, had been angled slightly away from the wall so that the device would 'jump' into the middle of the room. Fortunately, shortly before the explosion the hostess had noticed and pushed the bookcase back against the wall; she recalled hearing a 'clunk' but seeing nothing obvious, ignored it. It turned out that the device had toppled over, and instead of jumping into the room had 'jumped' back into the corner of the bookcase. Had it functioned as intended, it is doubtful if any of us would have survived.

The incident was widely reported, including in Time Magazine, back when it was a highly regarded international news magazine.

Later we also learned that the bomb had been planted by a trusted longtime domestic employee with perfect credentials. He had met every and all of the security criteria for employment, and had a reference file that would have guaranteed him employment by diplomats or senior executives for life; the best prospects for any local at the time.

After setting the bomb, he had been whisked across the border to safety in North Yemen. A few months later, after the Brits had left, he returned as a hero of the now ruling National Liberation Front. I never knew whether he was converted, coerced or cornered into his murderous act against people he knew and who trusted him.

I owe my life – on more than one occasion – to Arab friends, and sometimes to total strangers who have warned me, or extracted me from potentially lethal situations.

I remember one occasion when an employee of the refinery, whom I recognized but barely knew, approached me in the Little Aden souk (market) and advised me, "Go away, there are bad men in town." I took his advice. Later that day an expat was assassinated on the street in Little Aden.

I also know just how 'persuasive' radical revolutionaries can be, when seeking the cooperation of the unwilling, or the unconvinced.

Given the choice between your livelihood or your life, and the lives of your wife and kids, not surprisingly, alas, most will choose the former.

Which would you choose?

THE MA'ALLA STRAIGHT, ADEN

THE TOWN OF ADEN

Battle of Steamer Point

Aden, Armistice Day (Nov 11) 1967

"Abdul, bring me another Brandy!"

The shout, above the raucous cries of a flock of alarmed crows, echoed through the first floor apartment in a brief moment of silence before all was drowned again by the chatter of automatic weapons from the street below.

Six months later, from his armchair, the Colonel, a World War II veteran, listened intently to the stereophonic recording of the battle from his home in Chelmsford, England. "There's a chap who knows what he's doing," he said pointing at one of the speakers. "He's shooting at something. Listen: short bursts, not just wasting ammunition."

Steamer Point, Aden, South Yemen, was a duty-free port and one of the world's busiest when all the shipping through the Suez Canal stopped to shop and take on bunker fuel. The Six Day War in June 1967 closed the canal and Steamer Point died with the loss of its shipping lifeblood. (Not to mention the growing insurgency as local forces, egged on by regional and Cold War interests fought to claim the Spoils of the exhausted and dying Imperial 'Estate').

The handover to a local administration had been carefully prepared as a quasi democratic two chamber government.

An upper chamber made up of the traditional rulers, the tribal Sheiks, and a lower chamber of elected members, sustained by a local military and police force.

As the withdrawal progressed it became increasingly evident that the British handover government was no match for the revolutionary zeal of its communist competitors, the Maoist National Liberation Front (NFL), and the Soviet-Egyptian backed FLOSY (Front for the Liberation of Occupied South Yemen).

No sooner had the last troopship weighed anchor than the civil war erupted. The police and local army defected, many joining one or other of the rebel forces and the tribal rulers fled with as much booty as they could, leaving the NLF and FLOSY to duke it out. The civil war was over in a few days, the mainly urban commercial supporters of FLOSY, whose talents lay with the pen and the phone, proved no match for the illiterate Bedouin supporters of the NLF whose talents lay in the sword and the gun.

Eerily reminiscent of events in Afghanistan 53 years later . . .

By November 11th, 1967, the date of the battle, the 130-year British presence was just a few weeks away from its announced withdrawal by the end of the year. As the once sizable military base evacuated, its defensive perimeter shriveled to include only the airport at Khormaksar, and the foreshore at Steamer Point. As the troops withdrew, emboldened insurgents picked up the tempo of their attacks: the better to claim later that they had driven out the British Imperialists.

Those of us left behind were almost exclusively young men, told that we were "essential" to the operation of the BP refinery and the handful of other businesses that intended to remain after the troop withdrawal and the handover of power to a native administration. In fact, we were the "disposables," but too young and overpaid to realize that, as all the bigshots - along with wives and children - were evacuated before the turnover.

We, the "essentially disposables," were told that should things go wrong, we would be extracted by helicopter from a designated location with one bag of personal effects not exceeding 25 pounds, following a pre-arranged signal. In our case, the signal was a series of blasts on the refinery siren, upon which we had 20 minutes to get to the pick-up location. Supposedly there was an Aircraft Carrier just over the horizon to extract us should the need arise.

A quarter of a century later, I met a retired Royal Navy helicopter pilot who had made a successful business selling retirement lots in Cape Coral and Lehigh Acres to US service personnel stationed in Europe. It turned out that he had been serving on HMS Hermes in the Gulf of Aden in November 1967, at the time of the evacuation of the Aden base, and was one of the pilots tasked with our extraction. Needless to say, it provided justification to hoist a few beverages.

It is indeed a small world.

The Colonel:
Colonel Sir Hugh Boustead

Abu Dhabi 1972

"D'you play Polo?" the old man barked, slapping his boot with a riding crop and eyeing me with a mixture of suspicion and curiosity.

I knew what was happening. One didn't just drop into the Officer's Mess of the Trucial Oman Scouts in Abu Dhabi without a pretty good reason.

Fort Jahili
Trucial Oman Scouts, Abu Dhabi

I had just been introduced to the Colonel, Sir Hugh Boustead, who was to give away the bride at a wedding where I was the best man. The groom, Bill Arber, was an Oxford contemporary of mine, who had grown up in The Sudan, where his father was a career Foreign Service Officer, and the Colonel an old family friend.

The problem was, I had long hair. Not long by today's standards, but a 'Beatles' length haircut, long by the standards of the time, when military types had crew cuts in the States and Brits wore 'short-back-n-sides' and the Colonel intended to cut to the chase and in short order, find out what I was all about.

Determined to be respectful but not obsequious I replied: "I've ridden a few horses in my time, Sir." "Humhh!" he snorted, "You better come and have lunch then. Sunday. One o'clock. Sharp! And bring your bathing suit."

"Thank you very much, Sir," I said, trying not to be intimidated by this rapid-fire barrage of commands. "How will I find your place, Sir?" I enquired.

"Follow that silly new road to Al Ain," he replied, waving his swagger stick in a southerly direction. "When the road ends, just keep going" he harrumphed, slapping his leg with his riding crop to punctuate each command. "About 20 minutes, you'll see a big mountain: Jebel Akhdar they call it. Can't miss it. We're at the bottom of it. Bring those chaps over there," he said, waving his swagger stick towards a couple of uniformed officers wearing the black and white chequered headdress and black rope 'khutra' of the Trucial Oman Scouts.

Before the formation of the UAE, the seven Sheikdoms of Qatar, Abu Dhabi, Dubai, Sharjah, Fujira, Muscat and Oman were British Protectorates. They dated back to the collapse of the Turkish Empire after World War One, when piracy and raiding had once again become the order of the day in the region.
To protect their shipping and trade routes, the Imperial Brits imposed a 'truce' among warring tribes, and, of course, assumed responsibility for their 'protection.' Hence, the formation of the 'Trucial' Oman Scouts and the 'Protectorate,' as it was then called.

British Officers, in the tradition of T. E. Lawrence of Arabia, led Bedouin troops on camelback who patrolled the interior desert against raiders, where only irrelevant lines on maps defined territorial borders. It goes without saying that the recent discovery of vast oil reserves had generated intense national competition for previously uncontested tracts of desert!

The following Sunday morning, we set off in a jeep along the brand new four-lane median-divided highway which sliced through the ever-changing desert sandscape. We drove through barren pebble-strewn formations where the rock had been swept clean and polished smooth by wind and sand, except for the little tails streaming behind the pebbles where the wind had missed some sand in its haste.

Through sets of giant dunes, great waves of sand, hundreds of feet high, all clean and perfect, their razor sharp crests dividing light from shade as they snaked along the ridges, pausing with the wind for breath, as it were, in their relentless onward march.

It was a surreal drive. I don't remember seeing another vehicle on the no-expense-spared freeway that slashed its way through the eternal ever-changing desert.

GIANT SAND DUNES ON THE ROAD TO AL AIN

Every so often we passed a bewildered Bedouin with his family-laden camels halted at the curb, gazing at this unfamiliar intrusion into his routine migration, and no doubt questioning his previously unerring navigation skills, honed through generations.

STILL THE OFF ROAD VEHICLE OF CHOICE IN THE EARLY 1970S

When the road ended abruptly, and without warning, at the little oasis settlement of Al Ain, we kept on going as instructed. After twenty or so minutes of bouncing over the desert sands, the mountain emerged through the desert haze. Sure enough, down there at the base was a cluster of buildings, and before we crossed the final few hundred metres, there was the Colonel in his jodhpurs, wearing a brown Trilby hat and striding out to meet us.

TOS CHAP WITH JEBEL AKHDAR AHEAD

The Sunday Curry Lunch is an experience unknown to those unfortunates who were not privy to the Raj: the Imperial British regime radiating from India, and which sought to civilise the world . . . and did, at least for a generation or two. To this day, after three generations, cricket and bureaucracy are the only activities in which India and Pakistan compete without killing each other.

But I digress.

The Colonel had other ideas for us before indulging in his curry lunch. "First order of the day: exercise the horses." he barked, giving me a look that said 'now we'll see if you can ride a horse.'

I had no illusions. These were not your docile riding-stable ponies, but highly-strung thoroughbreds, mostly stallions with other things on their minds than taking tourists for a stroll in the desert.

The Colonel had known Sheikh Zayed, the ruler of Abu Dhabi, since the latter had been a six-year-old princeling, and in his retirement, had been entrusted with the care of the Sheikh's stable of magnificent Arabian horses.

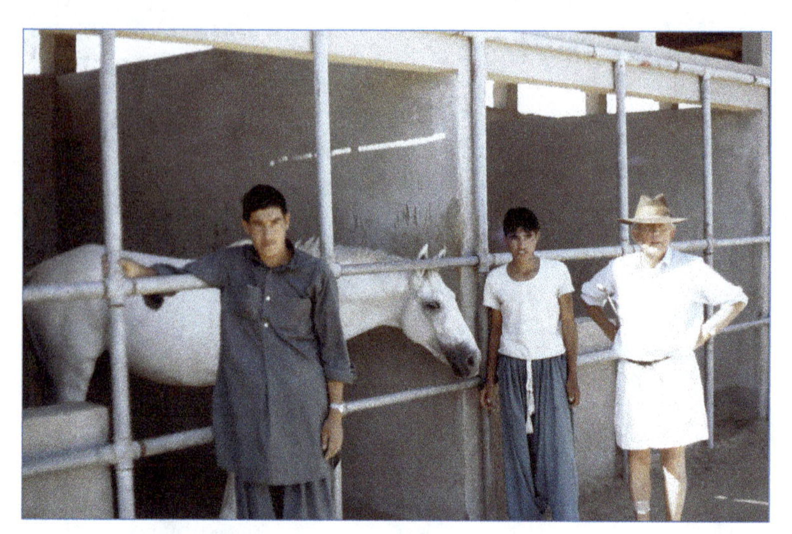

THE COLONEL WITH BALUCHI GROOMS AT ZAYED'S STABLES
with riding crop and brown trilby hat.

Before retiring to Abu Dhabi, The Colonel had been posted as Political Officer to the Aden Protectorate of South Yemen, as it was then known, and where I had just recently completed a three-year stint at the BP refinery in Aden. The role of a 'Political Officer' was one of great influence, without official rank or authority.

In regions where colonisation involving the establishment of a full government bureaucracy from Governor to Assistant District Officer – complete with subordinate police and military – was impractical or impossible, the Political Agent came into his own. Advisor, counsellor, and confidante of the Ruler, he was also the chief spy, influence peddler and Machiavellian Manipulator on behalf of his Imperial paymasters.

The Rulers of Muscat and Abu Dhabi had both recently been discreetly deposed in favour of their British-educated heirs, deemed more 'suitable' for the changing times. Sheik Zayed's predecessor, Sheikh Shakhbut,* in the saddle, so to speak, since the 1920s, had been deposed when Abu Dhabi's enormous oil reserves started to produce enormous royalty payments.

But back to the Colonel and his curry lunch.

Most of the buildings in the cluster were stables, apart from Boustead's modest bungalow and a small barracks that housed half a dozen grooms from Baluchistan. The Baluchis had our horses saddled up and waiting for us, so off we went. The Colonel headed for some big sand dunes about half a mile away, and we rode the horses up and down the dunes while the Colonel extolled the virtues of exercising horses this way.

Anyone who has ever tried to climb soft sand dunes knows how exhausting it is. I remember thinking of an old children's song that went:

> Oh! The grand old Duke of York
> He had ten thousand men
> He marched them up to the top of the hill
> And he marched them down again!

Obviously, a well established British military tradition.

It didn't take long to tire the horses, and we were back at the stable where the Baluchis took over, removing saddles and bridles and sponging down the horses while the Colonel, waving his riding crop, slapping his boot and shouting orders like: "Abdul! Jeeb wahid bag of oats ya silly old bugger", orchestrated the process as if conducting a symphony.

Returning to the bungalow, we were instructed to join him for a swim before lunch was served. The Colonel was immensely proud of his new swimming pool, which, for his generation, was the height of luxury. The pool was a 10ft x 20ft above-ground cement block rectangle, about four or five feet deep. At one end was a hose bib running continuously to compensate for the water seeping out between the blocks.

THE COLONEL IN HIS POOL

Still wearing his brown trilby, he 'walked' up and down in the pool doing a breaststroke with his arms, while explaining how important it was to wear a hat at all times, in order to avoid getting sunstroke and going mad. None of us were wearing a hat. He could have been the inspiration for Noel Coward's musical ditty ' . . .Mad dogs and Englishmen go out in the midday sun . . .'

The Colonel also expounded, and not too subtly, on why it was "unsporting," or at least inappropriate, to bring women into the desert. This was clearly directed at me and my companion, a beautiful brunette with legs that made men weak at the knees and women hiss with envy, and this in the days of the mini skirt that had not then, or yet, caught on in Arabia.

MY COMPANION ISA, CONSULTING WITH A CAMEL

Once we got into the curry lunch proper, we were able to get the Colonel into reminiscence mode, and the true dimension of the man's life and legend started to emerge. Born into an upper-crust family, he had gone to the right schools, and when World War One broke out, he signed up immediately, much against his family's wishes.

Aged 16 – too young for combat duty – he had been assigned to the Merchant Marine as a midshipman, languishing in the South Atlantic where, to use his words, "nothing was happening." Seeing a recruiting poster while on shore leave in Cape Town on his 17th birthday, he jumped ship, signing up under an assumed name as a private in the South African light Infantry "to have a crack at the Bosch," as he put it.

Six weeks later, he was in the trenches in France; listed as a deserter by the Navy, for which the penalty was the firing squad.
By the time the war ended, Boustead was a field-commissioned officer, an unprecedented feat for a private, especially a "colonial,' who had been wounded and decorated for gallantry on several occasions, and who was still wanted for desertion by the firing squad!

Through family and connections, he was able to circumvent a court martial and the firing squad with a 'King's Pardon'; re-assume his real name and get accepted at Worcester College Oxford, where, coincidentally, I matriculated forty-odd years later.

Soon tired of the society cocktail circuit, the future Colonel volunteered as a cavalry officer for the White Russians and went to fight the Bolsheviks in the Caucasus. Back in the UK after losing that one to the Reds, he was then shipped off to The Sudan, where he spent several years on camel-back, surveying and mapping the vast remote Western provinces of Kordofan, Darfur and Bahr al Ghazal. 15 years later, I would be navigating in Northern Darfur using the old Ordnance Survey maps bearing Boustead's name in the bottom right hand corner, among the six or so surveyors credited with creating the map.

It was during the Colonel's time in Sudan that he had become acquainted with Wilfred Thesiger, the eccentric British adventurer who, seeking the source of the desert locust, criss-crossed the Empty Quarter of the Arabian peninsula with the only tribe that could or would do it. In Thesiger's book 'Arabian Sands,' he describes walking barefoot for a day on the scorching desert sand, to earn the respect of the barefooted Bedouin who regarded him as a soft European incapable of suffering the hardships of their way of life. It cost him the skin off his feet, but earned him their respect as a fellow traveller!

The Colonel held Thesiger in high esteem, unlike Col. T.E. Laurence (of Arabia), with whom he was also acquainted, and who he dismissed as "a poser!... a glory hound!"

As Europe descended into Fascism in the '30s, and Mussolini invaded Ethiopia – both to avenge the Italian defeat at the battle of Adwa in 1896,** and to grab a piece of Africa for his Imperial ambitions – Boustead was assigned the job of creating the Sudanese Camel Corps, a mobile military force designed specifically for the vast empty regions of the Sudanese Sahil.

When the war came, he infiltrated his Camel Corps across the Sudanese border into the highlands of Tigray in Northern Ethiopia and mounted an extremely successful guerrilla campaign that drove the Italians out of Ethiopia, in what was arguably the first defeat of the Axis powers in the Second World War.

Fifty years later, in the dying embers of the Cold War, I was infiltrating Tigray in Northern Ethiopia along the same path, with Irish Nuns and bags of cash, in unofficial support of the rebel forces who successfully displaced the Soviet backed regime of Haile Miriam Mengistu.
It was not until some years later that I read his biography, 'Winds of the Morning' and realised that once again, I had trod, like 'Good King Wenceslas's page,' in Boustead's footsteps.

TRADITIONAL ARAB DHOW

Once sail powered, now increasingly motorised.

At the time, neighboring Dubai was an epicenter of the world gold market. At $32 an ounce, an estimated one third of the free world's annual gold production passed through Dubai, much of it smuggled into India, where it was worth $250 on the black market.

Known then as The Creek, Dubai was home to a small fleet of traditional Arab dhows: the sail powered wooden vessels that had plied the coasts of the Indian Ocean since the time of the Prophet.

Or so it seemed. Rumour had it that some of the dhows had been fitted with hydrofoils and twin-turbocharged Perkins diesels, that, when engaged, could lift the entire vessel out of the water and make sixty knots: more than enough to outrun the fastest Indian Coast Guard vessels!

Hardly palatial, Shakbut's palace" – where the cash had to be delivered – consisted of a few simple whitewashed adobe buildings inside a walled compound. As the flow of oil grew exponentially, so did the cash, which was soon being delivered in bales by the truckload, until every room in the palace was full of money. Only then did the old boy agree to put his cash in the bank on condition that he could look at it whenever he wished.

THREE GENERATIONS IN THE PALACE COURTYARD

for the Majlis, the traditional personal audience with the Ruler on the day preceding the Juma (Friday) gathering at the mosque. The youngster has not yet graduated to manhood, and the right to wear the Khunja, the distinctive dagger and tribal identity.

A banker friend who was transferred to Abu Dhabi after narrowly escaping an assassination attempt in Aden (a bullet passing through his cheek and out his mouth for the loss of a few teeth), organised the transfer of Shakhbut's cash. He described it thus: a pickup truck delivered the cash from the Palace in bales, dumping them on the sidewalk outside the little British Bank of the Middle East (BBME) on the corniche.

At the time, the BBME building was nothing more than a ground floor pillbox with bars on the windows and a BANK sign above the door: straight out of a Western movie. After dumping half a dozen bales on the pavement, the Land Rover pickup headed back to the palace for the next load, while a coolie with a wheelbarrow picked up the bales, one at a time, and trundled them into the bank. The millions lying unguarded on the street caused not a flutter of interest in the pedestrian traffic of Bedouin, such was the disdain for paper money at the time.

Footnotes:

*Sheik Shakhbut had been the ruler of Abu Dhabi for thirty years when its vast oil reserves were discovered in 1958. When oil and revenues started to flow in 1962, the old boy first insisted on being paid in gold. This was a problem for the concessionaire (BP) and after much haggling, the old Sheik agreed to accept cash . . . but it had to be delivered to his palace. No doubt he had been persuaded that he could always go down to the souk and buy gold with his cash, which, at that time, was still on the gold standard at $32 an ounce and in fact his neighbouring sheikdom Dubai was doing a brisk business smuggling gold into India for an 800% return if you did not get caught!)

**The Battle of Adwa (Sunday March 1st. 1896) in Tigray, Northern Ethiopia was the first and only time in the 19th century that a modern European army was defeated by an indigenous African army. An 8,000 man Italian Army was wiped out.

Such was the stain on the Italian national psyche, that in 1936 Mussolini set out to avenge this 40-year-old blot on the National Escutcheon by invading Ethiopia (Abyssinia, as it was then called) with all the monstrous weapons of the era, including aircraft and poison gas.

Despite a most eloquent and articulate appeal to the League of Nations by the young Emperor Haille Selassie, the inept organisation failed to act in accordance with its founding charter, thereby emboldening the aggressors – Italy, Germany and Japan – to precipitate World War II.

Ninety-four years later, in 1990, I was escorted through the battlefield by an Ethiopian whose grandfather had fought in that famous battle. He described in extraordinary detail the course of the battle, and the deployment of the various combatants, as the fortunes of the battle swung back and forth. Then, as he described the end game – when the tide of battle swung irreversibly in favour of the Ethiopians – my guide became quite emotional, describing how his grandfather had wept at the sight of so many good Christian men slain.

He was referring to the enemy, the Italians.

Ethiopia is a Christian nation, an island of Christendom in an ocean of Islam, to use their own characterization. Only in this context can one fully appreciate the significance of his sentiment.

CHAPTER 2:
Kenya and Tanzania (East Africa)

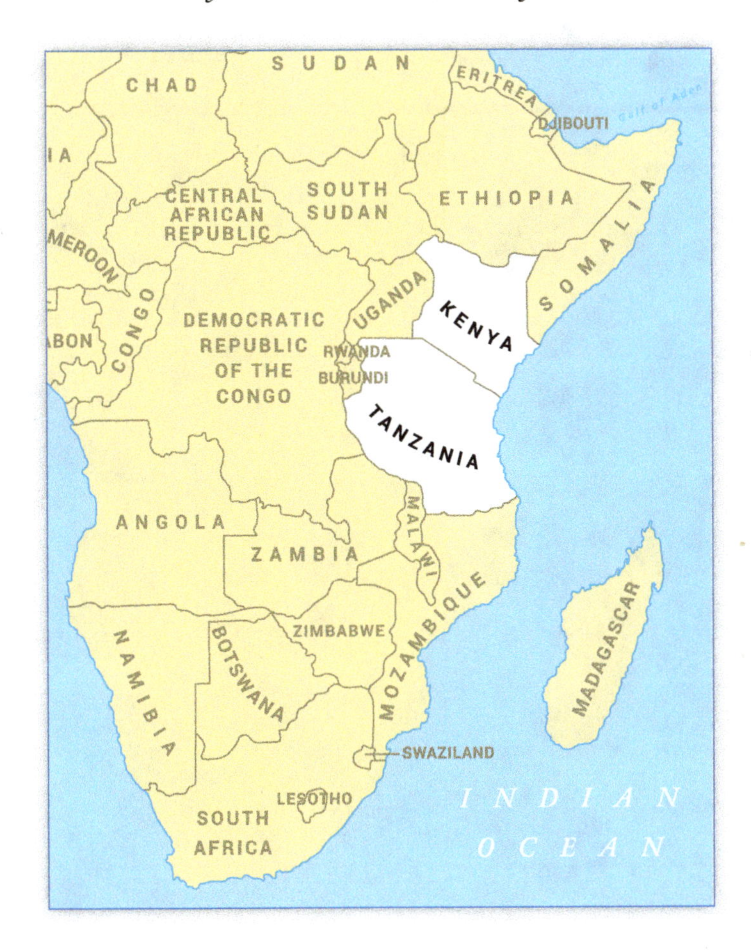

ENCOUNTER WITH BOADICEA

Manyara, Tanzania ,1969

"You weren't nervous, I hope", said Iain, peering back at me with his impish grin from the relative security of the driver's seat, as I stood trembling in the exposed open well of the LandRover pickup. I wanted to smite him but I knew I'd been had.

Indeed, I was not only nervous but still trembling from what was certainly, so I thought, a narrow escape from certain death.

We had just recently arrived from Wilson Field, the Bush pilot's Nairobi base and home of the Karen Blixen Bar, at the short dirt strip known as Manyara airport, in Peter Austin's Piper Pacer.

Peter's plane, a 160hp fabric tail-dragger of early 1930s design, was rated for about 650 lbs useful load, of which about 300 lbs was fuel. At 5500 ft altitude and 90°F with three big guys and some kit, we were way over gross . . . but somehow got off the ground – in defiance of both the laws of gravity and the Italian aeronautical genius Bernoulli – and made the two-hour trip to Manyara successfully.

I should add that on the return journey, from the short Manyara dirt strip just three weeks later, we were unable to break the bonds of gravity, and literally fell off the cliff, dropping a thousand feet or so, which happily, thanks to the Rift Valley was available, and so managed to inch our way back up to an altitude sufficient to cross the East wall of the rift and land safely back at Wilson field!

But I digress.

PETER'S PIPER PACER:
call sign India Whiskey

Back to the story, the Lake Manyara National Game Park lies at the base of the Western escarpment of the African Rift Valley, where the African continent is slowly splitting apart.
The procedure, in the absence of phones, radios and the like, was to fly down the wall of the escarpment and buzz the camp at the base of a small waterfall until we were sure that our arrival had been noticed. Then we would land at the Manyara strip, topside of the western escarpment, and await pickup, twenty or thirty minutes later.

All went according to plan, and Willie Arber and I climbed into the Land Rover pickup with our kit.

The Manyara airstrip, a red murham dirt strip, ran slightly uphill to a crown, and then downhill for about a hundred yards, to the edge of the escarpment, which fell off steeply to the Rift Valley floor some 2,000 ft. below. Unburdened from 400 lbs of Willy, baggage and me, Peter and his Piper Pacer – call sign India Whiskey – leapt off the ground and headed north back to Nairobi.

Willie had climbed into the cab with Iain and Oria taking up all the available seating so, with no other options, I volunteered to ride with the bags in the open well at the back.

And so it was that I met Boadicea.. Face to face you might say.

From the airstrip, a fifteen minute ride, via a tortuous series of hairpins down the escarpment, brought us to the entrance of the Manyara National Game Park, from where Iain's camp was perhaps another 20 minutes or so down what could best be described as a track: definitely not a road
The cab which enclosed Willie, Oria and Iain had a very small rear window, about eight inches wide and perhaps four inches deep, so I could see a bit of all of them but not much. No matters, it was exciting, the trees were full of baboons shrieking welcome or alarms, I could not tell. The vegetation was lush and there were impala, buck, Cape buffalo and such to excite a suit and tie commuter from a London office a few days prior.

I hung on as best I could while the truck lurched and bucked down the rutted trail

Suddenly the LandRover swerved off the track and crashed through the bushes into a clearing perhaps a hundred yards across, and stopped dead. In the clearing was a herd of twenty or more elephants who were clearly not pleased with our intrusion. In particular, and distinctly displeased, was a huge elephant with magnificent tusks confronting us not 30 yards away, ears flapping , trunk waving, shrieking, head tossing and pawing the ground in an unmistakable threat display.

Paralyzed, I froze, looking desperately through the tiny window at my three companions for some sign or assurance that it was ok. But no! They were all three also frozen and staring motionless ahead at the raging elephant.

Desperately, I looked around for escape options.

Lie down in the well maybe it won't notice you. No! Don't move!

Dash for the baobab tree 50 yards away. No! you'll never make it and the lowest branch is 20 feet up, and the bark is as smooth as marble and 15 feet around.

God help me! It will flip this truck over like a tiddlywink and I will be crushed under it. They're in the cab: they'll be ok. What can I do?

The entire herd was now shrieking, trumpeting, and milling around like cavalry preparing to charge. And then she charged.

Too late! Time to die.

I froze, staring at her as time stood still. I shall never forget that face... the second or two that the charge lasted was an eternity. Those massive tusks that could have tossed me like a rag doll . . . that incredible haughty stare as she looked down her nose at me, her head held high with a look of benign contempt.

And then she stopped, not ten feet away. Snorting, trumpeting, pawing the ground, ears flapping, waving her trunk as she blew dust in the air and then... backed up. But only to repeat the performance at least two or three times more. All the while the rest of the herd was carrying on, much the same, but at a discrete distance.

After a few more charges, she settled down and, along with the rest of her family, wandered off to be about their business.

As I stood in the back of the truck dripping with sweat, still desperately struggling with incontinence, Iain poked his head out the cab window, slowly turning his head to look up at me, and with that wicked, mischievous grin that only he can wear, said "you weren't nervous, were you?"

I wanted to smite him, but I knew I'd been had, and reluctantly had to concede defeat.

BOADICEA

As I later discovered, in his doctoral thesis on elephant behavior, "Boadicea" was central to his study of aggression. She was also supposedly predictable, so Iain took great delight in exposing his unsuspecting guests to her performance. Not all his guests took it as good clean fun: at least one – a celebrated author and environmentalist, who later became a close personal friend – who was not amused, and described the experience in a subsequent book.

Many years later in Haines City Florida, I was invited to a friend's house for dinner. The hostess, Sara-Beth White, was a talented artist. On the dining room wall, there was a spectacular painting of hers. It was a charging African elephant. The background was unfamiliar, but the tusks, the head held high, looking down her nose . . . there was no doubt.

"I know that elephant," I exclaimed!

"Don't be silly," said Sara Beth, "I painted it from a magazine photo."

"That is Boadicea," I insisted. "I have met her face to face, and she lives in Manyara, Tanzania."

A few days later Sara Beth called and said "You were right! I painted it from a photo in a National Geographic, and it is Boadicea!"

A Telephone call from Mto wa Mbu

Tanzania, January 1969 or so

"Telephone not working," said the man 20 feet away and seated comfortably on the ground, his back against a magnificent shade tree. He had been watching us for several minutes as we tried in vain to raise the operator on the other end of Mto wa Mbu's only telephone.

The phone was housed in a cast iron red kiosk with the letters VR under a crown prominently cast above the door. Once a common sight on street corners throughout England, this icon of Her Majesty's phone service, the public phone box (manufactured during the reign of Queen Victoria hence the monogram VR for Victoria Regina) had found its way somehow to this remote frontier of the former British Empire.

The Mto wa Mbu telephone

Mto wa Mbu, a village 17 miles from the entrance to Tanzania's Manyara National Park, was a strip of thatched huts and small cultivations of bananas, yams, cassavas and such, along an unpaved murham road; red dust in the dry season and red mud in the monsoon rains.

About midpoint was the telephone kiosk, outside a small shack which housed the operator waiting for someone to lift the handset and request a connection to number so-and-so.
This day there was no reply from an operator.

Now local protocol deemed it impolite to ask direct questions, especially of strangers. So the conversation went something like this, with the man sitting under the banyan tree: "It's a pity the telephone is not working" said Oria, nonchalantly.

"Yes, it happens sometimes, but mostly on Wednesday," replied the man sitting under the tree.

"Do you have much trouble with elephants eating your bananas?" asked Iain.

"Yes, before, but not now since they put bang-bangs around to make loud noise in the night, so they go away," replied the man.

"That is good to hear" said Iain, who had instigated the use of long-fuse firecrackers called 'flash-bangs,' to spook elephants who were raiding the banana plantations at night.

"When do you think the telephone might be working again?"

"Maybe when she finish her washing," replied the man.

Ah Ha! Now we had the first clue! More small talk about elephants, bananas and the weather, so as not to be rude and close the dialogue, and then: "By the way, where is a good place to do laundry around here?" asked Oria casually.

"The women go down there to the river," he said, pointing to a small bridge about a quarter mile down the road. "Ah! That is good to know," said Oria.

After a bit more small talk, we exchanged greetings, climbed into the Land Rover, and set off down the road towards the bridge. Stopping on the bridge, we looked over the rail, and there, a few feet below, were half a dozen local ladies who had each staked out a flat rock by the stream, and – armed with a box of Tide or Persil – were pounding their smalls by the water's edge.

"Jambo Masoori!" (Hello, how are you?)

"Masoori, Habari? (Well, and you?) came the replies.

We watched the local laundry in process for a few minutes, engaging in light banter about the womens' children, school, or cultivating issues, and then casually ventured "Is there a telephone in Mto wa Mbu?"

One of the ladies looked up and said, with authority "Telephone not working!"

Aha! We were getting close. How are the yams this year? Yams not ready till next month . . . and then, after an appropriate pause, "When do you think the telephone will be working?"

"Maybe when she finish her washing " said one of the chatty washers, indicating with a nod towards the woman who had first advised us that the phone was not working.

Now the riddle was answered, and the question was what to do!

Answer: 'patience' . . . so we watched patiently, with the aid of chit chat as the pile of washing diminished. We ascertained where the telephone lady lived: about a mile down the road, we estimated, since distance was measured in time – walking time that is – and often includes conversational pauses during chance encounters on the way.

"May we give you a lift?" we asked.

"Santi sana" (thank you), whereupon some other ladies remembered that they too lived in that direction, so there was a further delay while they finished up, and we four crammed into the front, while three ladies with large baskets of wet laundry squeezed into the back of a small battered LandRover, of the 109 series to be precise.

ORIA, IAIN AND THE BATTERED LAND ROVER
of the 109 series.

Mindful of the objective of the day's mission, which was to make a phone call to Nairobi, we managed to keep the telephone lady in our care by delivering the other two to their homes first.

At the telephone lady's home, it was soon apparent that the telephone would not be back in service until her laundry had been hung out to dry.

So we helped hang out the laundry.

"Could we give her a lift to the telephone since we were passing that way?" we asked politely.

"Yes, that would be good," came the reply. And so, two or three hours after the first attempt, we were now a few minutes away from success.

Back at the phone, the man was still resting comfortably against the tree, and he and the phone lady exchanged the normal pleasantries in Swahili. It was then that I broke protocol and almost blew it.

As she opened the door to the shack abutting the phone box, I said: "Please, can you get me Nairobi such-and -such a number?"

"NO!" she exclaimed, spinning around and glaring at me.

"Fuste you pick up de telephone," she said . . . and when I say 'Hello, can I help you?' THEN you give me the number."

Thoroughly chastened, I followed my orders to the letter, and after some further delays, while other operators further up the line were located or awakened, we succeeded in completing the phone call.

Before returning to the camp in Manyara we stopped to buy a large stick of bananas for "Sarah's teenage daughter." This had become a ritual for the weekly shopping trip to the Mto wa Mbu market, and I was soon to learn why. Inside the park, bumping our way down the rutted pathway in the open Land Rover to Iain's camp, there was suddenly an elephantine shriek and crashing noises in the bush close by, and a large but not-quite-full-grown elephant with one tusk emerged, and chased us down the trail, shrieking and waving her trunk.

"There she is," said Iain, stopping the truck for her to catch up and collect her bananas. Sarah's daughter would lie in wait, knowing somehow not only the sound of the Land Rover, but when it was on its way back from Mto wa Mbu with bananas.

They say an elephant never forgets, and I'm inclined to agree.

THE FAT LION IN A TREE

Manyara, Tanzania, 1969

"Watch this," said Alan, as we slowly approached the lioness draped over a tree limb ten feet off the ground. She was snoozing peacefully on a full belly, which was bulging like a great sack over the branch on which she was spread-eagled.

Slowly, we inched closer. She opened one eye. A little closer, and both eyes were open. One step closer, she lifted her head, lips curled and fangs bared; she let out a menacing growl. We froze.

Alan dropped a pebble from the handful he was carrying, and we slowly backed off, watching her head go down and eyes close as she resumed her nap.

A few minutes later, we approached her from a different angle and were met with exactly the same reaction. Alan dropped another pebble and we backed off.

After two or three more approaches from different angles, to my great relief, we called it a day.
I was fascinated by the process, its predictability, and the fact that she did not appear to build up any resentment to our serial interruptions of her postprandial snooze! That had been my fear all along.

Anthropocentrically projecting my instincts on the lady lion, I expected her to tire of our repeated intrusions and decide to get rid of us; with disastrous consequences for at least one of us, most likely me, being far less bush-savvy than the M'zungu* Alan. In other words I didn't get it, at least til the next day, when we returned to the tree.

The lioness was gone. Alan sought out his pebbles and paced out the distance from each to her position. Each pebble was exactly the same distance from where she had been. The pebbles defined the circumference of a perfect circle.

FAT LION IN A TREE . . .
sleeping off a good dinner

The circle defined her 'comfort zone:' outside it, no problem; inside it, you were in her space . . . watch out!

This was just one of a crash course in "bush-savvy" that Alan inflicted on me.

On another occasion, we had been observing a troop of baboons for an hour or two, safely downwind, while Alan explained their social hierarchy. Imitating their alarm calls, he could send them scampering to the safety of the trees, then after a discrete interval – while the baboons chattered away with their version of 'what's the problem' – he could give the 'all clear' bark and they would descend from the trees to the feeding grounds: the bold racing ahead to get to the choice areas, while the timid hung back to be on the safe side.

On the way back to camp, we came into a clearing, face to face with an equally surprised old Cape Buffalo bull, a notoriously irascible beast (the Swahili adjective is 'Kali' which means angry or bad tempered). I froze.

Not so Alan, who immediately charged straight at the old bull, waving his arms and yelling at the top of his lungs. Spooked, the two-thousand-pound beast turned and bolted.

"What on earth were you thinking?!" I asked, as he returned, wearing a self-satisfied grin. "Well," he said, "fight or flight are the first thoughts of most creatures when surprised; my brain works faster than his, so I helped him make up his mind!"

Alan was born and raised in Kenya. and spoke fluent Swahili. He had learned much of his bush craft during the Mau Mau rebellion, when many young men like him had 'gone to the bush' for months at a time, to track and hunt down the gangs that hamstrung cattle and committed unspeakable ritual atrocities on both white settlers and untold numbers of Africans who opposed them.

Now, still a few years away from the fame and fortune that would soon attend him as the foremost African Wildlife filmmaker of the time, he was a struggling wannabe living with his wife Joan out of an old Land Rover, following his subject 'beast(s)' until food or film ran out, and they had to return home to scrape up the resources to continue.

Alan Root was one of those larger-than-life characters who thrived in the great untamed outdoors, and would probably have been locked up had he been forced to conform to a civilized suburban environment.

Alain Zubair, visiting French photographer chatting with Oria, Iain Douglas Hamilton with Alan Root (white shirt) in the background. The Camp Building is upper right.

Totally fearless, especially around animals, he already bore impressive scars from 'close encounters of the animal kind'. Among these was what he called his 'tree frog hand.' The index finger and most of the flesh of his right hand was missing, courtesy of a puff adder bite that almost killed him. He nevertheless exhibited remarkable dexterity with the claw-like remnant.

He had come across the 'puffer', one of Africa's nastiest (nonpolitical) snakes. Short, fat and slow-moving except for its strike, it lies by the natural paths that small rodents use to travel between burrow, water and food, beating a pathway in the process.

Such paths attract more traffic, including people. Coiled in a muscular 'S,' the puffer lies in wait for its victim, its haemotoxic venom causing massive and fatal hemorrhaging in its prey. Alan had picked up the snake and displayed its finer points, including milking its venom for the benefit of his companion, who was taking photos. While she changed film, he released the snake and prevented it from escaping by pulling it back repeatedly by the tail.

Thoroughly irritated, the snake got its chance as Alan tried to grab it again behind the head, scoring a partial strike on his index finger. Miles from anywhere he took a shot of anti-venom, only to discover that he was among the one in three allergic to the antidote. By the time they got him to a hospital he was delirious. Thinking his delirium was due to the venom, the hospital gave him more anti-venom, which put him in a coma and almost killed him.

Root, typically recording everything on film, managed to get his companions to photograph the entire progression of the trauma, which he carried with him, labeled "Day 1, Day 2, Day 3 . . ." up to about Day 7, when the great black haematomas that had been his fingers and arm started to recede.

Alan's right buttock bore a perfect set of teeth marks from a leopard bite which he needed no encouragement to display at cocktail parties or in restaurants, it mattered not.

Scrambling around some rocks, he had spotted a fresh kill under a ledge, and jumped down to investigate. He had not, however, seen the leopard sleeping beside its kill, so landed in a full squat, his butt in the leopard's face.

"I don't know who was more surprised," said Alan, recounting the story, while pulling his pants back up. "We both took off like rockets in opposite directions, but I got to keep the scars."

The next time our paths crossed, some years later, he had an impressive six-inch scar on his right calf. Snorkeling in Mzima Springs to film hippos tiptoeing underwater like bulky ballerinas, he got between a large bull and the object of its affection. The hippo, perhaps blinded by love, narrowly missed taking Alan's head off. Its eight-inch tooth-tusks closing behind the glass of Alan's face mask crushed his eye-glasses, missing his face by a whisker.

As Alan wisely, for a change, turned to flee, the hippo grabbed his leg and tossed him like a rag doll onto the bank.

Fast forward a decade or so to 1989, Northern Uganda; I was assembling a ferry at Pakwach, just below the spectacular Murchison falls on the upper Nile. Below the falls, there reside a number of the world's largest and happiest crocodiles, feasting on a never-ending supply of fresh carcasses washed over the falls, and of course, the odd unwary native fisherman or washerwoman. You can tell how happy they are by the size of their smiles, as they bask on the banks between meals.

The new ferry at Pakwach below the Murchison Falls. Fred Cuny and Jean-Michel Goodstiker, in the white shirt.

Murchison falls, funneling the Nile into a chute!

And who should show up? None other than Alan Root.

Ensconced in a hide close by, he was filming a 20-foot mama croc guarding her nest. When I say "close by," I mean a few feet away. Get the picture? Since our last encounter, after the Mzima springs hippo event, he filled me in on his latest close encounter of the wildlife kind.

He had been filming gorillas in, as I recall, Rwanda. Armed with his brand new state-of-the-art camera, he was recording an 800 pound silverback – the biggest of gorilla males – playing in his nest with a juvenile. As usual, Alan was pushing the envelope to get a little closer when suddenly the juvenile spotted him and charged, beating its chest in the classic gorilla threat display.

Alan, in character, was unintimidated and kept on filming. The youngster, his threat display having failed, panicked and fled into the safety of grandpa's nest. Grandpa Silverback, like any good parent or grandparent, took up the challenge and charged.

"I knew I was in trouble," recalled Alan, "because he was horizontal." The chest-beating threat display is conducted upright, while the real deal charge is horizontal, knuckles on the ground and the torso horizontal.

Seizing Alan, the Gorilla picked him up, bit him in the side and threw him, still clutching his precious new camera, about thirty feet into some bushes. Having dealt with the problem and taught the juvenile a lesson by example, the old silverback ambled back to his nest to resume his nap, leaving Alan to limp back to camp with a couple of cracked ribs and an impressive tapestry of cuts and bruises.

I forgot to ask him what his life insurance premium was.

His African wildlife films collected multiple awards and were serialized in the UK by Anglia Television, under the banner "Survival Africa" and Alan's extraordinary life story was documented by CBS as "Lights,

Action, Africa," narrated by the late Ed Asner and in "Two in the Bush" on YouTube.

Iain and Oria Douglas-Hamilton, George Adamson, Alan and Joan Root – all of whom I was privileged to know – were, I believe, more influential in switching the culture from killing to conserving African wildlife than 99% of all the politicians and celebrities who ever jumped on the conservation bandwagon.

note:
***m'zungu** is Swahili for a European / white guy who is fluent in the language and culture and considered 'one of them'

The Lion King

Kora, Northern Kenya, Circa 1982

"Is that for Black 'n Tan if he gets out of line?" I asked. "Good God, man! That's for me," quipped George. For a moment, I was stunned, and then George's true meaning dawned on me. Of course, how silly of me.

We were standing over an open grave, next to two mounds in which were interred the remains of "Boy" and yet another lion, nicknamed Christian I think, both of whom George had had to shoot when they had attacked and/or tried to eat someone.

After a pause to regain my composure, I replied, "Not too soon I hope." I had missed the mark so totally with my comment. "Whenever" said George, with the kindly grin of an old sage trying to bring a young student up to speed.

Returning a lion raised in captivity to the wild is not easy. They must learn to kill to survive, a skill normally taught to most offspring by their mother. What starts as rough-and-tumble play as a kitten must morph into going for the jugular, and a quick kill to survive in the wild. Lions raised as pets are delightfully playful 20-pound kittens that grow into 50, 100, and eventually 450 pound cats with no fear of humans. Having bonded from infancy with their owners – and worse, associating them with food –the subtle distinction between 'with' food and 'is' food is easily confused in a hungry lion's mind.

Such was the mistake of the two buried lions. George's sidekick Tony Fitzjohn bore impressive scars on his throat, souvenirs from one of the buried lions. He and George had been doing a walkabout with some visitors, so the story goes, when they came across one of their former rescues that had been in the wild for some time. The lion recognized them and bounded over to greet them. Tony, never shy, especially with an audience, started to wrestle the beast. In no time, the lion had him on the ground and by the throat for the kill, leaving George no choice but to shoot the lion. "Am I going to die, George?" asked Tony weakly, in shock and bleeding profusely from his throat. "Probably" said George without a flicker of emotion, "but we'll do the best we can!"

Miraculously, Tony made a full recovery. His throat was terribly lacerated, but amazingly, the jugular, windpipe and spinal cord were not damaged. It is entirely possible that the lion was in fact just playing a little rough and had no intention of making the kill, but alas the luxury of reflection is not available in such moments of decision.

George Adamson had been a big game hunter who turned conservationist after being commissioned to kill a marauding lion that was preying on local livestock. After tracking down and killing the lion, he found to his horror that she was the mother of three young cubs. His late wife Joy's book 'Elsa the Lioness' – about raising one of these lion cubs and returning it to the wild – was an instant bestseller, and was later made into the blockbuster movie 'Born Free' with Virginia McKenna and Robert Holden starring as the Adamsons.

MESS HUT AT KAMPI YA SIMBA

Now in his late seventies, George was living in a camp consisting of a cluster of four or five 'tukuls,' round, palm thatched mud-and-wattle huts, and a 'mess' hut. The mess hut was rectangular, also palm thatched, but with no front wall. In it was a long table and some camp chairs, where meals were served and people assembled.
Above the table was a light bulb that could be hooked up to the Land Rover's 12-volt battery on special occasions. Otherwise, kerosene lamps were the order of the night. The entire complex was enclosed by a 5 foot chicken wire fence.

Water was delivered from the nearby Tana river in five-gallon cans. If the water carrier was running, which was rare, it was called running water. Either way, it was decanted into a couple of oil drums, one of which was perched on some bricks over a fire. Boiled, filtered and bottled it was safe to drink.

The toilet was a hundred yards or so downwind in a clearing outside the camp fence. It consisted of a slit trench over which were a couple of planks supporting an elephant's jawbone. It was completely open, there was nothing, nada, zilch between the user and the great African outdoors...

AN ELEPHANT'S JAWBONE TOILET

In this particular posthumous application, properly worn smooth with usage or sandpaper, the jawbone is extremely functional, especially for Westerners who are uncomfortable or incapable of assuming the full 'deep-knee squat' that most of the world's people adopt for their daily 'eliminations!'

George, his assistant Tony Fitzjohn, along with a cook and a couple of locals made up the camp complement. Visiting were Georgio Forno, the bush pilot; a very nervous little Italian filmmaker and his beautiful assistant, doing a documentary on prominent African wildlife conservationists; and yours truly, who had managed to get included on the trip, thanks to Iain and Oria's connections.

We had departed from Wilson field, Nairobi's general aviation airfield from which charter flights and bush pilots operated. The main feature of the airfield – besides its dirt strip – was the bar named after famous aviatrix Karen Blixen, whose propellor was mounted behind the bar. Here were quaffed great quantities of Tusker, arguably Kenya's favorite beer, and many tales were told, some taller than others, of adventures in the bush, derring-do and good old fashioned gossip of 'up country' goings-on.

Before the mobile phone, this particular bar was one of East Africa's major information hubs (or waterholes, to use the local term), where game can predictably be found. Safari guides, journalists, bush pilots, wildlife photographers, and conservationists all hung out there, and if your particular 'game' was not there, someone would know where he or she was and when to expect them back.

We departed Wilson with our pilot Georgio Forno, in his high-wing single engine Cessna, and flew northwest to Mount Kenya, at the base of which we landed on a private strip at the Prettyjohn's farm.

Sumptuous fare at the PJ farm

Here we spent the night and were sumptuously fed and watered before setting off in search of George Adamson's 'Campi ya Simba' or Lion Camp.

I say 'in search of' because about an hour into the flight, (and being a pilot I had insinuated myself up front, right seat), I noticed that Georgio was scanning the barren terrain below, as though searching for something.

"Have you been here before? " I asked casually. "Never," he replied nonchalantly. Trying not to convey the impression that I might be alarmed at this information I asked, "What kind of Navaids", pilot-speak for navigation aids, "do they have out here?," thinking that at least there would be some kind of primitive ADF direction-finding radio beacon. "Eyeballs man, eyeballs!" he replied, and a little impatiently, I thought.

Thinking that four eyes might be better than two, I asked what we were looking for. "Well," he said, "the Tana river should be up here somewhere, and we have to follow it until we come to a horseshoe bend, and then hope we can find his strip or the camp. We'll buzz the camp, and when we've got their attention, we'll land and hope they come pick us up. . ."

I started staring down at the barren scrubby terrain below. There was no sign of a river or any human habitation, but the wrinkled foothills had valleys that looked like they might carry streams if it rained. Gradually, these fingers coalesced into a river valley with a muddy brown stream meandering through the flat ground, as the foothills morphed into the plain. Slowly the stream widened, and as the terrain flattened it began to meander in serpentine loops. Sure enough, in one of the loops there was a cluster of huts in a clearing.

A couple of low passes and someone waved, pointing in a direction which in a mile or so led to a cleared landing strip with a primitive windsock. Taxiing back to an apron that looked like it was the place to park off the runway, we stopped by a tree on which a crude hand painted sign was nailed. The sign read : KAMPI YA SIMBA. Beware of Lion: Stay with the Plane. Or something to that effect. We did!

THE AIRPORT WAITING ROOM AT CAMP KORA
Northern Kenya.

A few minutes later, an old beat up Landrover of the 109 series, with its windshield folded down, came bouncing through the trees and took us to the Camp. There, we were allocated our sleeping quarters, a hut with a folding camp bed, and then assembled back at the mess hut for a welcome cup of tea, followed by a guided tour of the neighborhood.

Outside the six-foot chicken wire security fence, someone – usually George or one of his assistants – carried the World War I .303 Lee Enfield that enhanced but did not guarantee our survival prospects. There wasn't much threat from big game except lion, because the surrounding terrain did not support much except dik dik and other small antelope and game: consequently, lion tended to be hungry. The real danger were the 'shifta', bands of poachers from neighboring Somalia who increasingly were armed with Kalashnikovs, and having slaughtered almost everything with a market value (whether horns, tusks or spotted fur), had started robbing and raping any unfortunates whom they encountered. Needless to say, white folk were especially prized in the exercise of their profession. "Bloody shiftas!" George would mutter under his breath whenever they were mentioned.

At the time, there was a young lioness with cubs that had not been long in the wild. She had mated with a wild lion called Black and Tan and would bring her cubs to the camp to show them off and collect a snack from George.

LIONESS WITH CUBS AT THE FENCE

The Tana River was a muddy stream that weaved its sluggish way towards the Indian Ocean many miles to the east. Perhaps a hundred yards wide at this early stage in its life, it was neither in flood from the peak of the monsoon rains, nor reduced to the pathetic trickle of the dry season, but somewhere in between. Nonetheless, its sluggish yellow brown waters provided ideal habitat for lethal predators ranging in size from crocodiles to the seemingly innocuous snail which hosts the deadly bilharzia worm.

While I was standing on the bank contemplating these things, the exuberant Tony – in the exercise of some primordial territorial instinct – pushed me off the bank into the river! I got out, if anything faster than I fell in, and once safely back on the bank, shook like a dog, raising a laugh from the group. And the tour continued.

At the time, George was trying his hand at returning leopard to the wild, something no one had been able to do successfully. The native population had been all but wiped out by poachers, their skins being highly prized in some cultures. Leopard skin slippers are high status symbols in Sudan, for example, as I learned a few years later when trying to decline the gift of a pair without causing offense. (In neighboring Zaire, President Sese Seku Mobutu never appeared anywhere without his leopard skin hat).

TONY WITH A LEOPARD CUB

So the next stop on our tour was the leopard compound, where George had two young specimens secured, to protect them from their mortal enemy the baboon, which will hunt down and kill a young leopard without mercy.

Once fully grown, a leopard can take care of itself and will occasionally take down a single baboon if it strays outside the security of the troop. But generally, leopards prefer less dangerous food, like dik dik for an appetizer, or a Thompson gazelle for a main meal.

The two leopard cubs had reportedly been bought from an exotic pet shop in Paris and smuggled into Kenya by sympathetic Sabena pilots, then out to Kora all 'under the radar' as they say. Ironically, the laws forbidding owning or trading endangered species apply equally to catching them in the wild and selling them to pet shops or buying them from pet shops and returning them to the wild! And since the law does not discriminate, the program was conducted 'hors de loi!'

Then there was the lioness, turned loose a year or so previously, who had just reappeared with cubs. She had taken to showing up at the camp with her new family, for what might be called food aid. This meant that George's budget, lean at the best of times, now included buying a cow or camel haunch every week or so.

The other major topic of conversation, and of more concern than the growing dependence of the lioness on food aid, was her mate; a magnificent wild lion called Black 'n Tan for his distinctive dark mane and contrasting tan color. Like all good male lions, he was very territorial and had taken to wandering around the camp at night, roaring to warn off intruders. Worse still, there were reports that he had been following, if not stalking, people.

This was a major concern of the moment, since George did not want to lose any of his helpers and certainly did not want to lose another lion! It was with this information fresh in my mind that we arrived at the grave site and I made the mistake of assuming it had been prepared for Black 'n Tan.

REGULAR DINNER GUEST
last weeks remains in the background

TEA TIME WITH GEORGE ADAMSON

Back at the camp, it was time for a cup of tea and the afternoon siesta before assembling in the mess hut for supper. Conversation over supper was fascinating, centered on the shifta and general gossip. Newcomers to the scene were particularly entertained by the various 'local' residents who had adapted to the feeding ritual in the mess hut. An assortment of birds and small mammals, squirrels, mongoose and gerbil-type creatures would appear on the table, pause briefly to check out their options, seize a choice morsel from someone's plate and disappear with their booty as fast as they had appeared. The initial surprise soon gave way to hilarity as one guest after another was robbed!

GEORGE WITH ANOTHER REGULAR DINNER GUEST

After dinner, as the sun set, we adjourned to an outdoor table with 'picnic' chairs for a nightcap. Not far away was a rock outcrop worn smooth by the ages. It formed a great dark bald head of granite, a smooth rounded skyline backlit by the last remnants of daylight.

On it was the unmistakable silhouette of a pride of lion enjoying the view and the belly warming rock as the night cool descended. It was magic.

LION SILHOUETTES

AFTER DINNER NIGHTCAP UNDER A FULL MOON.

I had been well briefed on the protocols, and had brought with me a large bottle of Johnny Walker Red, George's favorite beverage. He was too proud and gracious to be so crass as to charge his visitor/ guests, so the form was to ask if one could make a donation to the cause, and ignoring his dismissive rejection of the proposal, lay down, unobtrusively, an envelope which of course contained a contribution! It was a lovely and refined ritual allowing each to give discreetly, according to his means. The whisky was a personal gift which I had been assured would be much appreciated, and it was.
Nocturnal sounds are different to those in the day, but they are aplenty. The mewing bark of a distant hyena, the chatter of birds and chirping of crickets . . . and from time to time, the unmistakable sound of Black and Tan as he patrolled his turf. It wasn't the full roar of the MGM movie lion, but more a sort of mewling roar followed by a series of huffs or coughs, but unmistakable in its message.

Later in the night, as the level in the whisky bottle ebbed, we witnessed a total eclipse of the moon. It was a complete surprise to all present: without newspapers or radio, who would have known? As the shadow crept across the moon's shining face, the light dimmed, and as the light dimmed, so did the chatter of the beasts and birds. There was an ominous portent in the air that we all felt as the silence descended with the darkness. At the totality of the eclipse, it was pitch dark, the moon just a faint reddish orb and the silence palpable. It was easy to understand how in earlier times such events were regarded with foreboding. The apocalyptic prophecy "and the moon shall be turned to blood . . ." comes to mind.

After a minute or two, during which it seemed the world held its breath, the first glint of light broke out as the shadow crept across the moon's face. Slowly, the light spilled back, driving out the dark, and with it, the sounds of life returned, chasing away the silence.

After the celestial show, most of the company turned in, but George and I decided to pay another tribute to Mr. Walker's fine product. Some time later, I received the call of nature and wandered out of camp the hundred yards or so, to the place of the elephant's jawbone. There

I sat, feeling no pain, my pants around my ankles, when I heard the unmistakable 'cough' of a lion.

Freezing motionless, I slowly turned my head to look over my right shoulder, and there, to my horror, not ten yards away, was Black and Tan eyeing me, so it seemed, like a choice morsel on the menu.

Such moments tend to loosen the bowels, but it was too late for that. Neither flight nor fight were options, and once I stopped grasping at straws for salvation, strangely I felt a great sense of relief. I remember thinking "don't move" and "don't make eye contact" and then, reconciled to my helplessness, thinking: 'If this is it, so be it, at least my friends will get a laugh. "He was eaten by a lion while taking a crap in the bush! Dumb sh*t!!" '

So I sat there motionless, for what seemed like eternity... concentrating with every fiber of my being on being unappetizing, while Black and Tan made up his mind.

Slowly he walked around me, making his huffing sounds. At one point, he stopped not twenty feet away directly between me and the camp, turned his head and stared straight at me, twitching his tail in the manner of cats with things on their mind.

Then – swishing his tail – he wandered off. Waiting till he was out of sight, I wasted no time finishing my business and returning to my whisky. "I thought we'd lost you," said George, puffing on his pipe between sips of whisky. When I explained what had happened, he listened unmoved, looked at me, finished his whisky and said: "Well done: it would be a shame to lose him. Goodnight. See you in the morning." And off he went to his hut, puffing his pipe.

What did he say, I thought? "It would be a shame to lose him." He meant Black and Tan, not me!

And he was right. I consoled my wounded ego with the notion that it was a backhanded compliment. I had done my part well, and we had both survived. With that thought, I slept well.

The next day after breakfast, the Italian filmmaker took some pictures and chatted with George about his project, while Tony, Georgio and I went off to check on the leopard cubs and gossip about mutual acquaintances: who was where and doing what and with whom, etc. All good grist for the bush telegraph when we got back to the bar at Wilson Field.

When it finally came time to leave for Nairobi, we all piled into the Land Rover and bounced our way down the trail to the airstrip. As we emerged from the trees onto the strip, there was the plane . . . and standing on top of the cabin was a magnificent cheetah. Cheetahs love to take the high ground when looking for a likely snack. Alas, startled by our appearance, he took off into the woods before I could get a photo.

Once we were all on board, we bid our farewells and Georgio fired up the bird and taxied to the downwind end of the runway. A quick run up: mag check, feather check, controls free and clear, then open the throttle and we started the takeoff roll. As we gathered speed, I glanced out the window and there was the cheetah alongside, chasing us down the runway, not ten yards away and keeping up with us 'till we lifted off! What he would have done with us had he caught us we'll never know . . .

Seven or eight years later, we had a total lunar eclipse on Sanibel Island, Florida – where I lived at the time – and I sat outside watching the whole event. It was only the second eclipse that I had seen in my life, and needless to say, it brought back memories as vividly as if they had been yesterday.

It was also the day George Adamson died.

The next day, his death was widely reported in the news. He had been killed by a gang of Somali poachers. The story goes that a group of visitors, on their way from the camp to the airstrip, had been ambushed. Hearing shots, George and his driver, whose name I think was David, set off to investigate. Recognising instantly what was happening, they drove right into the gang. Even at 84, with his World War I bolt action .303 Lee Enfield, George was still a formidable shot and apparently nailed a couple of the 'bloody buggers,' as he called them, before they cut him down with their AKs. The gang fled, and their hostages escaped without being robbed, raped or worse.

George was buried in that very same grave, next to his two favorite lions, and with full honors from the Kenyan Government.

According to his biographer, the locals say that the lions come in the evening and lie on the mound of his grave. And I, for one, believe them.

Chapter 3:

Darfur, Sudan

EAT THE INDIAN MANGOES

Darfur, Sudan. 1985

"Excuse me Sir, there's a phone call for you from Geneva," said the waiter at the Steak & Ale restaurant in Key Biscayne, Florida.

I took the call from a complete stranger, agreeing to meet him at Miami airport the next morning. The caller represented a Sudanese-American enterprise that had just been awarded a major USAid contract and – desperate to recruit help with Middle Eastern language and cultural experience – had been given my home number on Sanibel Island by a mutual friend.

My wife Lorretta had given the caller the phone number of my host on Key Biscayne, whose wife had sent us boys off to the Steak & Ale, preferring to stay at home, and so was able to provide the phone number there.

Fortunately he had a long enough layover in Geneva to track me down.

And so it was that after a brief interview the next morning, between his connecting flights, I was hired for a year in Darfur as part of Uncle Sam's famine relief program.

En route to Khartoum, I stopped in the UK to visit my aging parents. They suggested I meet with an old friend of the family who had been a career civil servant in The Sudan during its 50+ years under British administration. Arrangements were made and at the appointed hour, we arrived at the home of Douglas McJanet, a former Governor of Sudan, for a glass of sherry and a chat, overlooking the 2nd green of the Woodbridge Golf Club.

After an appropriate time for social chitchat and a couple of glasses of sherry, I asked what advice McJanet could give to prepare me for the task ahead. He paused, looked at his drink, then at his shoe, and then – looking me straight in the eye – said "It's been thirty years since I left . . . things may have changed". . . and after another pause, "but I doubt it."

Then after yet another pause to contemplate his shoe and his drink, he said, "Two things I can tell you for certain. First: don't expect anything to work, never did and never will; and second: eat the Indian mangoes, not the local variety . . . much better!"

Thus briefed, I embarked for Khartoum a couple of days later.

Mid-flight we were diverted to Cairo because KHT Airport was closed. No explanation was given.

At the Cairo Airport, we were held in a very basic holding area: a room with no windows that was several decades overdue for fresh paint. It had that musty aroma that suggested neither broom nor mop had disturbed the dust since Cleopatra had tried nursing an asp.

We were watched over by some very disinterested uniformed guards who, likewise, were long overdue for clean, if not new, uniforms and showed no interest in anything other than their next cigarette break. Somehow, we managed to establish that Khartoum Airport was closed "TFN" because of a coup attempt.
This was not encouraging news since without Egyptian visas, there was no escape from our holding pen, until Khartoum reopened, or the airline agreed to fly us back to London.

Sometime the next day, tired, disheveled and still separated from toothbrush and toiletries, we were loaded onto an EgyptAir flight (known as Misery Air after its Arabic name Misr Airlines), and made it successfully to Khartoum, where we were advised that our luggage would arrive later 'Insha'alla.' This ubiquitous Arabic phrase, meaning 'God willing,' is applied to every event in the future.

Automated baggage handling had not yet penetrated this part of the world and separation of passenger from luggage was the norm, so the form was to go home and send a driver to the airport every day, to check for the conjunction of God's will and the arrival of one's luggage.

As it happened, I arrived in Khartoum during Ramadan, the month of fasting and the Hadj, when devout Muslims journey to Mecca to pray at the Kaaba, the black cube where Abraham was stayed by divine intervention from sacrificing his son Isaac, and where the Prophet Mohammed, after his flight from Medina, regrouped and launched the Islamic faith.

This may seem an irrelevant digression into history, but in fact it has some relevance to foreigners arriving at Khartoum during Ramadan.
Airplanes have greatly simplified (and multiplied) the annual pilgrimage to Mecca, the Hadj.
Khartoum was many days away from the Red Sea by camel, followed by a risky ride on a sailing dhow to the port of Jeddah and a further journey through the desert to Mecca.

Now, thanks to Boeing and Airbus, the faithful are whisked to Mecca in a few hours. This has made Khartoum somewhat of a Mecca (no pun intended) for the 'Hajis' of East Africa . . . and they are legion.

Few of the faithful owned suitcases, or even the now ubiquitous backpacks. Most packed their traveling necessities in a 'kiis,' pronounced keys with emphasis on the 's' as in hiss, the colloquial Arabic word for a sack. Without an automated baggage handling system, luggage was off-loaded manually and piled up on the tarmac outside the airport terminal, called 'the baggage claim area': a giant pyramid of suitcases, boxes, sacks and 'kiis', the ubiquitous plastic tote resembling a large version of the grocery bag you get when opting for plastic over paper at the supermarket checkout.

Climbing all over this mountain of luggage were young boys, known irreverently as baggage monkeys. Circling around the base of the heap were passengers shouting and gesticulating at the boys, who clambered across the pile, lifting up and waving bags, tossing back rejects and burrowing for others, until they got the nod and a tip earned for uniting a traveler with his luggage.

I remember marveling at the chaotic efficiency of this simple unplanned and unmanaged process and comparing it with the fully automated baggage handling that we take for granted. Exercise, employment, suspense –yes! – and anxiety . . . will he find my bag? Excitement of the hunt . . . yes! he found it! . . . fulfillment, and the reward for success: the tip. Have we really progressed that far?

Arriving in Khartoum, I spent a few days getting familiar with the Arkel-Talab head office procedures and personnel, before being thrown into the field. For preparation, I was sent for a week to Kosti, the capital of Kordofan and the Arkel-Talab distribution hub for that province, where I was initiated into the mysteries of large-scale famine relief distribution logistics.

The railway, our supply line, passed through Kosti on its way to Nyala, at the end of the line, in the remote Western province of Darfur.
Darfur means 'the house' or 'home of the Fur,' a wild and rebellious people, with a long history of resistance to central authority. It had been the wellspring of the Mahdi's rebellion in the 1880s which had annihilated the Anglo-Egyptian garrison in Khartoum, beheading its commander General (Chinese) Gordon.

Even before departing Kosti for Nyala, it was apparent that – true to form, the Fur were acting up. A relief convoy had been ambushed and hijacked in a remote mountainous corner of Darfur. Needless to say, this concerned the USAid folk, who confronted the local Fur Chief, demanding that something be done.

The Chief, clearly embarrassed, assured them it would not happen again. Not satisfied with his assurances, the Aid guys insisted on knowing specifically what actions the Chief had taken, actions he was clearly reluctant to reveal.

Eventually caving in to their demands, a couple of Aid guys were escorted to a site on a remote mountain road, beside which were six fresh graves.

On top of each grave was a severed human head.

There were no more questions, and it didn't happen again. . . as the Chief had promised.

Welcome to the world of consequences . . . !

Reel in the Monkey!

Dafur, Sudan, 1985

On the inside of the front door of the two-bedroom trailer that was home, office and Regional HQ of Uncle Sam's famine relief activities in Darfur, was a large sign. It read: 'REEL IN THE MONKEY' in big bold letters; right in the face of anyone contemplating opening the door to go outside.

At the other end of the room was a hand-cranked winch that wound in (or out) about 100 feet of steel cable that could be locked with a ratchet. Outside, on the end of the cable, was a young male baboon. Now for those who have never owned one, a little background might be in order. Although monkeys, baboons are not the cute and cuddly variety. Males are about 40 lbs. of testosterone and teeth, permanently united with a VERY bad attitude! They also have bright red and blue hairless bottoms, which other baboons find very attractive.

By reeling him in, and confining him to a five foot radius, it was possible to leave the house without being torn to shreds; otherwise it was not!

In the wild, baboons live in troops: groups of several dozen, in an extremely hierarchical social structure dominated by one, two, or occasionally, three males. From puberty onwards, the life of a male baboon is about fighting; battling through the competition on the road to success and domination.

The females move straight to the inner circle, to groom and bestow their favors on the dominant male(s), which of course is why the young males compete so fiercely to move up in the hierarchy.

Baboons are very smart and very dumb at the same time. Clever at finding water in arid places, they are very fond of salt, a combination which the Sudanese pastoral nomads use to trap them.

Baboon

yawning.. or something

Nomads place a block of salt in the hollow of a baobab tree, behind a slot just wide enough to insert an open hand but too narrow to pull out a closed fist. The baboon will follow the salt trail to the trap; reach in to grab the salt; then will not let go of the prize to get his hand out, thereby trapping himself. After tying the baboon up for a day or two with plenty of salt, and then releasing it, his captors are then led by what is now a thirst-crazed animal to the nearest waterhole.

Our monkey was a gift from one of the several hundred truck drivers who delivered relief supplies to eighty-odd distribution centers we supplied all over Darfur.

We had been experiencing a rash of burglaries; intruders having, on several occasions, gotten into the house at night and made off with briefcases and other decoys designed to thwart thieves from the real prize, which was payroll cash.

The traditional security system is a ten-foot-high wall around the compound, with an equally high gate manned by a 'Ghaffir' (pronounced gaffeer). The Ghafir is typically an elderly man, well known and respected in the community, who carries a good stick and lies most of the time inside the gate on a wood-framed hammock called a rakuba. The stick, ostensibly to repel intruders, is often used to enhance his own mobility, especially as he gets older.

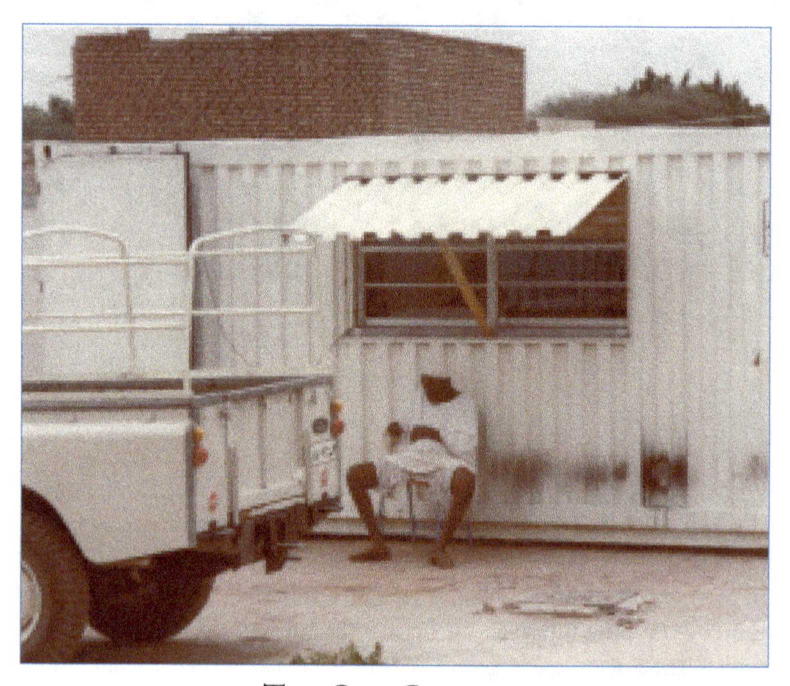

THE OLD GHAFFIR
in the exercise of his custodial duties

The theory goes that in a culture where age is venerated, the Ghaffir's turf is respected, even by the Haraami, as thieves are called. The practice also serves as a sort of social security, since the old boy has a bed, and a roof of sorts over his head, more for shade than shelter from the rare rain showers. He gets fed, along with the cook from the main house kitchen, and collects a small stipend.
Haraami, which technically means 'sinner' in Arabic, (and strangely

stems from the same root as 'Hurma', a woman, or its better known plural 'Harem'), are usually teenage boys, young and agile who ply their trade stark naked with a liberal covering of oil or grease.

This makes them very difficult to catch, even if one does wake up during a robbery. The good news is that because they rely on stealth and slipperiness to escape, there is seldom any violence!

Before installing the monkey, we had had an increasing number of nocturnal intrusions, although the Ghaffir insisted that he never slept during the night watch. This, despite the fact that many times in the wee hours one could peek out the door to see (and hear) him snoring away on his rakuba.

However, he was a nice old man, eager to please, and there was nothing to be gained by humiliating him with accusations, however justified. So one night, while he slept, my side-kick Ronnie and I crept over to the Ghaffir's rakuba, and – rather than wake him – took his sandals and stick from beside him, and hid them out of sight behind the house.

A few hours later, we were awakened by frantic banging on the door and shouts of "Haraami! Haraami!" Armed with flashlights, and clad only in minimal sleeping attire, we rushed out to search for the miscreants, still not sure what had triggered the alarm. "They stole my shoes and my stick," the old Ghaffir wailed, as we walked around the compound shining the flashlight into dark corners.

"From your feet?" I asked, trying to sound incredulous.

"La! La! Jamb er rakuba, (No! No! By my bed) "... and my stick too!" he cried.

"You didn't see them? So close to your bed?" I couldn't resist winding him up a little tighter.

"They are very clever, these Haraami, so fast you cannot see them" he whined.
"Even when you are fully awake?" I asked. This set him off again

"Abadan!" Never! "I never sleep on night watch", he insisted.
"Perhaps it was a djinn" I offered, seeming to accept his vehement assertions. The djinn or genie is a mischievous spirit that inhabits the dust devils, small convective whirlwinds that pop up in the heat of the day.

"Yimkin" (meaning 'maybe'), he replied, enthusiastically, seizing the chance to escape from the corner of mendacity in which he found himself. At about this moment, the beam of my flashlight happened on the missing shoes and wooden staff. "Allahuakbar!" (God is great!) he shouted, grabbing his stolen property.

Then after carefully reinstalling his feet in their sandals, and grasping the stick in his right hand, he straightened up and looked me in the eye, "Shoof keyf" he said triumphantly, "You see! I was so fast to raise the alarm that the Haraami ran away before they even had time to steal my shoes and my stick . . . so I could not have been asleep!"

Faced with such irrefutable logic, what could I say? I smiled, "Laazim (must be so)," I said, and we all went back to bed. Ronnie and I went to sleep and the old Ghaffir... .? Well, probably, but we'll never know.

One thing was very clear: the old Ghaffir was no match for the Haraami at night, but we still needed a gatekeeper to announce and such during the day, so for a time we reinforced the night shift first, with a policeman, and later, added a uniformed soldier armed with an Ak47. Nevertheless the haraami still managed to breach our security occasionally, and it was clearly only a matter of time before they found the 'motherload'.

Despite the policeman, the soldier with his AK47 and the old Ghaffir, the haraami still managed to penetrate the defenses. Whether by stealth, or complicity with the guards, or when they were all asleep, we would never know.

Until we got the baboon, that is.

The policeman and the soldier both quit immediately, but not the old Ghaffir. Faced with the choice between his money and his life, he chose the money. Retreating into a remote corner of the compound just beyond the reach of the baboon at full stretch, he rode out the night shift unburdened from his responsibilities. I don't know for certain, but I doubt if he slept much.

I do know one thing though: never again did we have an intruder. NOTHING came into that compound when the monkey was out and about.

A Riot for Belafonte

Nyala, Sudan 1985

"Not now! Can't you see I have guests?!"

Thus spoke the Nyala Chief of Police when I rushed up to him hot, sweaty and breathless, begging urgently for help, my hand-held radio crackling with panicky voices. He was standing in line with half a dozen other immaculately dressed senior officials - polished shoes and crisply starched uniforms with knife-edge creases - as a VIP charter plane taxied towards the 'red carpet' receiving line.

The VIP aboard was Harry Belafonte, along with an entourage of international journalists. Behind me, barely a quarter mile away, was a full-scale riot, as hundreds of hungry Sudanese stormed the flimsy stockade like the Zulus at Rorke's Drift, overwhelming the handful of police guards, and looting the supplies.

Harry Belafonte
at the Nyala Railway.

I was the Darfur Field Manager based in Nyala for Arkel Talab, a joint US- Sudanese logistics enterprise responsible for the in-country transport and delivery, to stricken areas, of a million tons of USAID famine relief (mostly Kansas-grown sorghum). Of this total, about 160,000 tons was destined for Nyala, for distribution to 12 initial hubs which ultimately fed 80 distribution centres. The grain arrived in 20,000 ton bulk carriers to Port Sudan on the Red Sea, and was discharged pneumatically in a few hours, to form mountains of grain on the dock.

USAID HELPING HANDS LOGO

From there, hundreds of labourers - working shifts around the clock - filled, weighed and sewed up 50 kg sacks emblazoned with the USAID 'helping hands' logo, for shipment by truck, train or donkey cart (typically all three) to a final destination, while folks back home were treated to stories of the 'mountains of grain rotting on the docks while people starve' variety.

Nyala is a dusty little town at the end of a rickety narrow-gauge railway built by the British in 1916 to move troops to the area, lest the Germans encroach on the British Empire.

In 1986, it was the distribution hub for Darfur - Sudan's Western Province and an area the size of France - with a total of 80 kilometres of asphalt road.

The contents of the irregular trainloads had to be offloaded in 24 hours to avoid huge demurrage charges, and were stacked in the railyard behind a flimsy fence guarded by a handful of police armed with night sticks. From here, it would be loaded into trucks and delivered to some 80 villages and distribution points as and when trucks and fuel were available.

As the drought worsened, more and more hungry squatters moved into town looking for food, and the railyard, having no security fence, was inevitably the destination of choice. (Spillage from torn sacks could be scavenged off the ground or from emptied railcars).

By the time of the Belafonte visit, this squatter camp had grown to perhaps 5,000, and it was increasingly obvious that it was only a matter of time before the dam burst, so to speak, and hungry hoards would be storming the rickety fence, overwhelming the handful of guards and police with their nightsticks. So it came to pass that when the chiefs of the police, army and civic affairs were all in the VIP receiving line, the dam did finally burst . . . a remarkable coincidence!

The hungry hoards swept over the fences like the Zulus at Rorke's Drift, overwhelming the panicked guards and police, who then started lobbing tear gas canisters around while in full flight. I too fled, to seek help, and save my skin, in equal measure, which explains why I was in such dishevelled condition when I reached the police chief.

By now, Belafonte had deplaned and was advancing down the red carpet to the reception line, while a second plane disgorged the press corps, many of whom were seasoned veterans of conflict, and - not surprisingly - at the first whiff of tear gas, sensed a good story. So, with nostrils flared, they set off towards the nearby railway yard, and the epicentre of the action.

AFTERMATH OF THE RIOT

It was at about this point that the Chief of Police grasped the gravity of the situation, not to mention the consequences to his reputation if all the food aid was stolen under his nose in the presence of an international celebrity, and so summoned the necessary reinforcement to save the day.

Apparently, Belafonte was intrigued by the events, and I was invited to join him at the home of one of the local dignitaries for an evening meal, an event that, under the prevailing conditions, was distinctly more social than gastronomic.

We sat on a carpeted floor around a 'tabah', the big metal tray on which communal food is served, breaking off pieces of the flat pancake of bread for dipping into the various dishes on the tabah, and always with the right hand only. There was a lively discussion covering a wide range of African affairs, and I was impressed by Belafonte's grasp of the issues and problems. He was well informed and genuinely interested; a refreshing change from the usual stream of journalists and celebrity voyeurs there for the photo opps.

The following morning, we finished the guided tour that the previous day's riots had interrupted, before departing with his entourage for Khartoum: his companions were anxious to get airborne before the heat-triggered turbulence made flight too bumpy. To my surprise, Belafonte invited me to visit him when I was next in New York.

Some months later, I took him up on his invitation and enjoyed a most pleasant evening of drinks and conversation in his New York apartment. Once again, I was impressed with Belafonte's grasp of Africa's problems; his genuine concern; and the realism with which he explored solutions.

Although our paths never crossed again, twenty years later in the Bahamas, at The Tingum Village Hotel in Dunmore town, Harbour Island – known locally as Ma Ruby's – I got to know the Percentie Family. Ma Ruby's husband, Humphrey Percentie, was a local icon: a songwriter and balladeer.

It was, in fact, Harry Belafonte who brought many of Humphrey's songs to the world. Among others, the Banana Boat song - with its refrain 'Day-Oh! Daaaaay-Oh! Daylight come and I want to go home' - was, I am told, a Percentie original.

True or not, oftimes it takes two to tango.

TRADITIONAL SUDANESE HOSPITALITY
dining on the carpet

Quicksands, Waterholes, Opportunists, and Others

Darfur, Sudan 1986

Harry Belafonte was one of the few visitors to Nyala who ventured to the front line where the action was. Most earned their stipends attending briefings from the press offices of the power centers, or at the bar of the Meridian Hotel, the gossip center of the International community in Khartoum at the time.

Before the Internet, every capital city in the developing world had a gathering point, a sort of information hub, where resident expatriates, journalists, and savvy folk gathered to gossip and glean information. In Aden, in the 60's it was the Crescent Hotel, in Nairobi in the 70's, it was the Lord Delamere bar at the Norfolk Hotel; in Khartoum, Sudan in the '80's, the Meridian.

Many of those who did venture down to visit were there for a photo 'op' and got out of there ASAP.

I remember one story of a well known televangelist, who shall remain nameless, and who was clearly uncomfortable holding emaciated babies in the heat and the flies of reality, while his photographer clicked away trying to catch empathy that wasn't there, turning to an assistant and growling: "get me to the Goddamn Hilton!"

WITH HIS EXCELLENCY, GENERAL SUWAR AL DAHAB
President of Sudan.

To be fair, some pretty good folk did show up trying to help, or tell the story as it was, or at least as they saw it.

Along with Belafonte, among the best were the President of Sudan, General Suwar al-Dahab, and a couple of guys from the Baltimore Sun: a journalist and a photographer who came down for a couple of weeks to document what we actually did.

After a couple of days, the Baltimore Sun guys had seen all there was to see in Nyala and wanted to follow the aid to its destination. This involved persuading a trucker to take them on board, which was accomplished with the aid of a cash benefit or some extra fuel, I forget which.

As I recall, we hitchhiked them on a truck heading to Geneina – on the Chad border – where there was a sizable refugee camp. It was not far away, a couple of hundred kilometers or so at most, but the so-called road should have been called an obstacle course. The route included deep sand, mud-puddled swamps and numerous wadis where, like thin ice, the surface could collapse into quicksand, and all without AAA roadside rescue.

QUICKSAND CONSUMING A TWENTY TON BERLIET TRUCK.
No problem for traditional transport (rear).

A few days later, they staggered back: dirty, disheveled and much the wiser, while their journalist colleagues in Khartoum "suffered" in air conditioned hotels, struggling to meet deadlines, and crafting spectacular attention-grabbing headlines to justify their salaries.

A shout out to our intrepid reporters, whose first hand experiences appeared later as the cover story of the Baltimore Sun's Sunday color supplement.

THE ROAD TO GENEINA:
Ronnie Austin deep in contemplation

And then, surprise surprise, there were the others. 'Opportunists' would be the kindest generalization: political, commercial, and ideological.

Among the more interesting of the political opportunists were the Libyans. Muammar Ghadafi was something of an international pariah at the time (mid 1980s) having endorsed the PLO; bombed a West German nightclub frequented by US servicemen; and challenged the US Navy in the Gulf of Sidi – which cost him two of his new Soviet MiG-29 fighter jets.

Perhaps to improve his image he joined the Aid band-wagon by sending a sizable convoy of trucks loaded with food aid across the border into Sudan, where they showed up in El Fasher, the capital of Darfur province and the seat of the Military Governor.

We had a distribution hub in el Fasher which required my attention once or twice a month.

Our weekly supply plane from Khartoum would fly to Nyala then El Fasher and back to KHT, or the other way round so I could hitch rides back and forth between Nyala and el Fasher to avoid the day-long drive over a bumpy unpaved road.

From time to time we would be instructed to deliver to a new destination often up near the Libyan border.

On one such occasion, on a hunch I had invited the governor to a plane ride under the pretext of seeking his advice reconnoitering the best supply route. The General loved the plane ride and so periodically we had to intrude on his 'busy' schedule for a reconnaissance flight . . . for which the General was always available.

When Ghadafi's convoy showed up in El Fasher, the Governor with whom I now had a useful relationship, threw a welcome Garden Party at his Mansion to which I was invited.

At the reception, I was engaged by a young Libyan fellow who spoke excellent English with a pronounced Birmingham accent, which was unusual to say the least. It transpired that he had studied electronics at Birmingham University in the UK.

He was very interested in the electronic equipment at the local airports, radar, directional beacons etc. and invited me to see the convoy camped about a mile outside the city.

The next morning I showed up at the Libyan Camp, where the 20-odd trucks were parked in a defensive circle like a Western wagon train.

The Camp was manned entirely by young men with short haircuts who were being ordered about by a slightly older man carrying an automatic rifle. Being escorted about by my new Libyan friend with the Birmingham accent, I noticed a truck stacked with sacks of powdered milk which wasn't covered by a tarpaulin. Protruding slightly between the sacks were metal tubes that could easily have been mistaken for the flash eliminators on anti aircraft guns.

I started to take some photographs, and no sooner had I raised my camera than the chap with the gun came rushing over yelling at the top of his voice "MAFI SOORA! (NO PHOTO!)

He grabbed my camera, and was about to smash it when my new friend with the Birmingham accent persuaded him to spare the camera and take the film, which he did, ripping it out aggressively!

Later that day, I got on the shortwave radio, our connection with the main office in Khartoum and mentioned the incident, without making a big deal of it. First thing next morning, a charter plane arrived, with an Embassy chap who introduced himself, handing me a business card with his name and 'Office of Humanitarian Services' for identification. He was extremely interested in the convoy and very disappointed that I did not have any photos for him.

Despite many visits to the Embassy I was never able to identify The Office of Humanitarian Services, or corroborate its existence. A chimera indeed!

There were a couple of other clues as to what the Libyans were up to. It was very hard to hire trucks for deliveries to Kutum and Um Kedada, two settlements up near the Libyan border: nobody would go there for love or money, and then suddenly everyone wanted to go there. Finally, I asked one of my regular truckers what the deal was.

"Backhaul," he said. "What's the backhaul?" I asked, still puzzled. "Electronic goods", he replied, and then the penny dropped. He wasn't talking about radios or VCRs: which don't need 8- or 10-ton trucks to haul. He was talking about Surface to Air Missiles (SAMs) and anti aircraft weapons for the rebels in the South.

A couple of years later, I was flying from Entebbe, Uganda into Juba in Southern Sudan in a giant C130 Hercules Cargo Lifter, at 16,000 ft. to stay above the SAM-2 operational limit, and the pieces of the puzzle came together.

No Flies on my Man

Nyala, Darfur Sudan 1985

An early indicator of a famine is a surplus of meat, along with rapidly falling prices in the local markets. This is pretty logical. As a drought intensifies, grass dies and livestock begin to starve, so farmers and herdsmen try to sell livestock before the animals starve to death.

And so it was in Darfur: meat was cheap. Mohammed the cook did all the shopping and we ate well . . . lots of meat. Not always the choicest cuts, but the charcoal barbecue was busy every night, and we didn't enquire too closely about what, exactly, was on the grill. It was always lamb or beef, which sounds better than goat or camel but honestly, with enough A1 barbecue sauce, tastes much the same.

Being in the logistics business, we had a chartered plane that– every two weeks – delivered cash for wages and 'essential supplies' for my sidekick Ronnie and me. Essential supplies included such luxury items as the aforementioned barbecue sauce; canned fruit; canned soups; mustard; ketchup; and other unobtainable delicacies . . . which may explain why the other expats from MSF, Red Cross, Save the Children and a host of UN-prefaced entities often dropped by at supper time to discuss urgent business!

Anyway, we used a lot of barbecue sauce while serving the NGO community, in addition to supplying the local populace with 160,000 tons of Kansas-grown red sorghum, compliments of Uncle Sam.

One day, returning from somewhere, I drove into the meat market on the outskirts of town. Now this is not quite the same as your local deli, with its choice cuts hygienically shrink-wrapped for display. Not exactly. This meat market was outdoors, in a patch of desert a sensible distance from town, for reasons that soon became apparent. FLIES! Millions and millions of them.

The Islamic ritual for preparing meat for the table is called halal. The animal is bled to death by cutting the jugular vein in its throat, before butchering the corpse into choice cuts for the cook. It is quick, relatively painless and very messy. The animal's life blood soaks into the sand, attracting swarms of flies.

Typically, there are a dozen or so meat vendors, each set up behind a table from which to display and sell his wares. Some are under one of the few shade trees, while others are under umbrellas, thatch or canvas shades. The rest, either late-comers or struggling start-ups, cope under the blazing sun. Tethered close by is 'inventory' still on the hoof.

The fresh-cut meat on the table is covered with a shimmering, blue-green iridescent blanket of food-frenzied flies. Hovering above is a huge black cloud of more flies, buzzing angrily for their turn.

Periodically, the butcher – seemingly oblivious, or simply resigned to the cloud – flaps a fan over the table, often at the approach of a customer, causing a volcanic eruption of flies, and showing the red of the meat briefly, before the dark, glistening blanket descends again.

Sickened at the realization that for the last several weeks, if not months, I had been eating this fly-blown worm-infested meat, I continued home, convinced that my days were now numbered and it was only a matter of time before I succumbed to some ghastly parasite-induced malady.

At my first opportunity, I confronted Mohammed and – as casually as I could – enquired as to his source for our meat. He confirmed my worst fears: it was indeed the dreaded place. Trying to hide my alarm, I asked whether he knew of a supplier with no flies. "My man no flies," he assured me, to my skeptical relief. "Could he show me?" I asked.

On the following market day, we set off, Mohammed and I, to the meat market. Weaving our way through the buzzing clouds that punctuated our trail, we threaded our way between the fly-smothered tables until we came finally to a vendor, standing behind a table with freshly butchered blood red meat, and miracles of miracles . . . not a fly to be seen!

Astonished, I stared at the fly-free meat, and at Mohamed, who was grinning with pride and self-justification, then back at the fly-free meat, speechless. As I stood there trying to fathom the incomprehensible, our butcher pulled out a can of Pif Paf from under his table and laid down a thick film of this lethal insecticide over the bright red meat, and for good measure into the surrounding air above and around, sending any and all hapless flies in his airspace spiraling instantly into the ground.

What can I say ? "Those whom the Gods wish to punish they grant their every wish." Perhaps.

THE UBIQUITOUS PIF PAF

Thirty-plus years later, I am still alive and none the wiser as to how anyone can eat such a quantity of a lethal nerve agent insecticide and survive. I can only assume that Mohammed, who was incapable of undercooking anything, must have successfully burned off all the toxic chemicals in the exercise of his culinary talents.

Follow-Up Footnote:

Thirteen years later, following the 1998 bombing of the US Embassies in Kenya and Tanzania, President Clinton authorized a cruise missile strike on the Al Shifa Chemical Factory In North Khartoum.

At the time, Osama bin Laden had been expelled from his native Saudi Arabia, been granted residency in Khartoum, Sudan and was reputed to be a major investor/owner of the plant. (See "A One-Armed Ethiopian named Abahdi.")

Rumor has it that the justifying evidence for the cruise missile strikes was gathered by Embassy intelligence chaps from soil samples gleaned from around the Al Shifa plant. They had allegedly identified traces of parathion, the insecticide ingredient of Pif Paf but also the prime ingredient of the far more lethal biological warfare agents Sarin and Tabun.

One of the justifications for Gulf War II was the WMD allegation that included the German-built plant at Samaria, Iraq, which manufactured enormous quantities of 'agricultural pesticides' – far in excess of any legitimate agricultural requirements – and the biological (warfare derivative) tabun likely was used in the massacre of over six thousand souls by Saddam Hussein's buddy 'Chemical Ali' in the Northern Iraqi Kurdish village of Halabcha.

I Am Not Competent

Nyala, Darfur Sudan 1985

"I am not competent to judge Sharia Law, " said the well-spoken Sudanese gentleman seated next to me, as he raised his glass and took a sip of the milky liquid it contained. "All my life," he continued, "I studied English Common Law, and then one day they said: you will now judge Sharia law!"

He paused, and looking straight at me, raised his glass again. "I am not competent to judge Sharia Law," he repeated.

There was a sincerity about his words that engulfed me in a tsunami of relief; and I believed him. You see I was 'hawajja,' meaning a foreigner, especially a white man or European.
We were here today and gone tomorrow.
We came with money and technology and good intentions.
We lived apart.
We were treated with courtesy, patience and tolerance, but we did not belong, and soon would go home to be replaced by a new face.

So we were like replaceable parts of a machine; not members of the community. We were components of a system, not friends or neighbors. It was thus unusual to be confronted by four local gents – strangers – who had stopped by for a social visit, and, weirder still, that they came bearing dangerous gifts.

We sat around the table in a double-wide trailer that served as both home and office. Just me and four Sudanese gentlemen who had made this surprise visit, dressed in traditional white robes and bulky white turbans.

SUDANESE IN TRADITIONAL GARB
white robes and bulky turbans.

As I soon discovered, they were the Chief of Police, the Mayor, the Judge and the richest and most powerful merchant in Nyala.

On the table in front of me were a bottle of Johnny Walker whiskey and a bottle of Beefeater Gin, gifts from the four unexpected visitors for which I thanked them profusely but did not dare touch.

You see, the Government in Khartoum had recently been overthrown in a military coup backed by the "Achwaanee" – better known as the Muslim Brotherhood – and Sharia Law had been imposed.

Alcohol is 'mamnooa' (forbidden) under Sharia, and the penalty is forty lashes outside the Mosque after Friday prayers, so the possibility that I was being set up was foremost on my mind.

Mohammed the cook had furnished us with glasses and the customary water and sodas when the gentleman to my left, who I now knew was the chief judge, produced a medicine bottle from under his robes, decanted a healthy pour into his glass and added a splash of water. The 'medicine' produced a milky white liquid on contact with the water exactly like Greek ouzo or Turkish arak.

"What is your medicine for?" I asked innocently. "Oh no!" he exclaimed, smiling. "This is Araki" referring to the local 'moonshine' alcohol made from dates. Somewhat taken aback by his candor, I asked, as casually as I could, "Is that a problem with Sharia?"

It was then that he made his unforgettable
"I am not competent. . ." statement that began this tale.

When he finished, I looked across the table at the other three, all of whom were smiling at me with a medicinal bottle of their favorite alcoholic beverage out in the open on the table in front of them. So! What had made these "heavies" of the community decide to drop in with such precious and dangerous gifts?

As the evening wore on, it became apparent.

The Nyala power station was relatively new, part of some foreign aid grant, but it only operated for a few hours at a time, a few times a month, and without warning. The plant had been designed to burn 'heavy oil,' a high wax content low grade fuel that was so viscous that it could not be pumped until it was first heated. The fuel was supposedly the 'bottom of the barrel' (the low value part) of substantial oil reserves that Chevron had found in Southern Sudan.

Ongoing civil conflict in the area, together with a host of common third world political and financial constraints on development, had meant that the particular fuel source was still years away when the power station was completed. Worse still, even if an alternative source had been available, there were no heated rail tankers with which to supply it; nor the funds to buy and operate them. So the power station served to maintain the water pressure by pumping up the water tower every week or so as needed and little else; burning expensive and precious diesel, as little as possible and only when absolutely necessary.

That was everyone else's problem, not mine. We had our own generator. A huge, extremely thirsty 350 KW Caterpillar, that ran 24/7 and consumed a couple of barrels – 80 odd gallons – of diesel a day.

THE 'BEAST' IN ITS CAGE
(far right)

We had acquired this beast cheap from some drilling rig or other that didn't have enough camels to drag it home, and even with every appliance running wide open, we could only use somewhere north of 10 kw and south of 20 kw.

To make matters worse, diesels do not behave well on light loads. They are sloppy, noisy, inefficient and tend to rattle themselves to death, which is expensive . . . and catastrophic when you live in a double-wide trailer that becomes an oven in minutes without air conditioning.

The solution was to add 'load.' So we hooked up the neighborhood with the proviso that each house would install an isolation switch, costing less than $10, so that when the main power came on – without warning of course – it would not blow up our generator.

At the time, I had no idea that we had 'wired' the most powerful men in the town.

By including us on their evening social visitation rounds, they were not only including me socially, but more importantly, letting me know that 'they had my back' so to speak.

From that time on, everything, 'for some reason,' started to run smoothly. Negotiations ran smoothly, issues with the local authorities were quickly and easily resolved, even those ugly confrontations so common in commercial disputes just melted away.

I remember one incident when a local 'contractor' with a grievance had been threatening me with those "we know where you live", and "you went to the market yesterday" thinly-veiled threats that are hard to counter except with 'Lebanese insurance'.*

Then one day, out of the blue, a stranger stopped me and told me that the contractor had left town and would not be coming back. No explanation was offered or needed, but I had the distinct impression that he had not left of his own accord.

Amazing how random acts of generosity, even when selfishly inspired, can return great rewards. I often wonder if I owe the successful renewal of my contract not to my own competence but to the self-serving interest of preserving my generator...

* During the Lebanese Civil War, 1975 and onward, when assassins were available at $50 a pop, it was not uncommon for prominent individuals to set aside appropriate funds to ensure that should they have an unfortunate accident, those parties that might benefit therefrom would also have unfortunate accidents... hence the term 'Lebanese Insurance,' a scaled-down localized version of the prevailing Mutually Assured Destruction (MAD) deterrence principle that happily preserved humanity throughout the Cold War... not to mention a few Lebanese businessmen!

A Fish Out of Water

Nyala, Darfur, Sudan 1985

From time to time, various uninvited strangers showed up at the door – well, actually the front gate – where my ever-vigilant Ghaffir held them at bay, pending my approval or otherwise.

One day, a distraught young man from the UK showed up with a tale of woe that did not surprise me but who I couldn't turn away.

He was a volunteer, an apprentice CPA who had temporarily quit his job to volunteer with Bob Geldof's Band Aid.

Eager to save the World and fired up with media stories of 'profiteering', 'corruption', and. 'Mountains of grain rotting on the wharfs while people starved,' he had chosen to travel by rail to save money, a noble motive.

The Sudanese railway, apart from running on rails, bore no resemblance to even the most crowded commuter trains of jolly old England... but nobody had told him that.

TRAIN ARRIVING AT NYALA

Cheap seats on the roof...

The trip from Khartoum took several days. The train had minimal facilities of the sanitation kind, nor ventilation, other than windows for those who had purchased a seat, like my destitute visitor.

Those traveling on the roof had no facilities of any kind, so not surprisingly relieved themselves as necessary, an unpleasant surprise for passengers who had secured a seat by an open window.

While he was sleeping, his briefcase had disappeared and with it, all his essentials: passport, money, and – of great concern – landsat maps of the region which he had promised, on his word of honor, to return . . . and no one in the crowded carriage knew or saw anything.

Happily, we were able to clean him up, cheer him up and feed him, all of which he needed badly, before packing him off back to Khartoum a few days later on our weekly charter flight. We kept in touch, and over time became friends.

He returned to the UK before my contract was up and wanted to meet me in London on my way back to the USA. "There's someone I want you to meet," he said.

So I did; contact him, that is, with my return flight schedule.
He was there at Heathrow as I disembarked, a taxicab waiting, which was all very nice. And off we went to London, to an address in the Kings Road, the trendy street of the day, where all the high-end boutiques, restaurants and celebrities lived.

We arrived at a fancy residence and I was ushered in and introduced to its occupant, none other than Bob Geldof...

For those who weren't sentient in the 1980s, Geldof and his Boomtown Rats, a minor league Irish Punk Rock group, started Band Aid, a charity rock concert for famine relief, which raised tens of millions, and propelled Geldof overnight into international celebrity status.

The ability to raise money does not, alas, come with the ability to spend it wisely or effectively, especially when managed by the celebrity and legislative classes unfamiliar with the practicalities of the real world.

A case in point involved a junkyard of 96 old trucks which we, Arkel-Talab, the in-country logistics contractor for USAid, had declined to buy for $600,000. A few months later a "Charity Organisation" bought it for $3 million.

I remember watching the closing of the deal on an NPR News clip, the same Irish broker that we had declined, shaking hands and congratulating the buyer with the words: "Best of luck! From what I know of Sudan, you're going to need it!"

Six months later only 6 of the 96 trucks were on the road, thwarted by lack of spare parts and the fact that semis and articulated trucks don't work off-road.. where 95% of the demand lay... I guess nobody told them that!

To be fair, it wasn't only enthusiastic amateurs who made the mistake of assuming that what worked back home would work in the Third World, now known as Less Developed Countries.

The UK provided a bunch of brand new Bedford 10-ton trucks to an NGO as part of an Aid contribution. Designed for level paved roads, they lasted for a few weeks off-road where the torquing from uneven surfaces popped the cabs off their mounts and broke driveshafts.

In a matter of weeks they were all abandoned.

Meanwhile, back in the King's Road, London, I was tired, irascible and unsympathetic to the idea that locals could be coerced into working for free, or expected to risk their all – their trucks – for little or no gain. Geldorf was courteous, gracious and listened.

Then he said something like "I'm a 'fookin' musician. You seem to know what you're doin' and I'd like you to run the fookin' thing for me."

It was a mountain too tall for me to climb at that moment, so I declined.

A decade or two later, on Harbour Island, in the Bahamas, I was enjoying an adult beverage on the veranda at the Harbour Lounge, a favorite haunt. On the next table, was a family gathered for dinner. The main guy looked familiar. It was Bob Geldof.

So I went over and, excusing the intrusion, asked him "are you still in the Band Aid business?" He looked at me for a moment and then said: "Are you still in the 'fookin' trucking business?"
God bless him!

The Apocalypse Now Gang

Darfur, Sudan 1986

"Why are you doing this?"
"What did you find?"
"Did you find gold?"
Asked the Sudanese merchant friend, with a knowing wink that said: 'Hey! my friend, you can share the secret with me.'

He looked at me with a kindly, bemused grin as I tried to explain why we continued to airlift supplies in the giant twin-rotor helicopters known in the military as Chinooks, weeks after the rains had ended and the routes were again open to trucks.

The fact that we would spend in excess of $5,000 an hour, hauling two tons of supplies a distance of under 90 miles, when we could deliver 10 tons by truck for under $100 just didn't add up. The more I tried to explain how U.S. government contracts worked, and that we had a three month contract with Columbia Helicopters out of Seattle Washington, the more he just grinned and wagged his head.

There was just no way that this shrewd old Sudanese businessman, one of the richest and most powerful merchants in Darfur, was going to buy into that one. Something else HAD to be going on!

The Sahil, as the semi-arid belt between the Sahara desert and the rain fed regions to the south is known, catches the tail end of summer monsoon rains, most of which is dumped on the hump, the mile-high East African plateau, and feeds three of Africa's four great rivers: the Nile, the Zambezi and the Congo.

BOEING B-107 VERTOL REFUELING

Not a lot of rain makes it over the 'hump' but usually enough for subsistence farmers to plant and harvest enough 'aysh' for a crop; next year's seed, and sometimes, a small reserve.
In drought years, crops fail. Drought in consecutive years leads to famine, and when the drought breaks, things get worse.

The rain is often abnormally heavy, causing flash floods which isolate large areas from relief. Even if seed is available, and not eaten in desperation, starving people and their emaciated animals are in no condition to plow and plant. And even if they are, it takes months before there's a crop to harvest.

A FLASH FLOOD:
blue sky, rain miles away, & hours ago

1985 was no exception. When the rains came, not only did they break the drought, but they washed out the railway - our fragile supply line - too. The road and rail tracks across the Sudan run east to west across the north-south gradient of the land, with the high ground to the north. When it does, the rain falls on the high ground. The runoff feeds down the mountain valleys into a series of wadis, or dry river beds which channel the water into flash floods, sweeping away everything in their path.

A CASUALTY, CAUGHT IN A FLASH FLOOD

To make matters worse, in the dry season, the desert winds move the finest particles first, so that the surface gradually acquires a 'skin' of fine powdery dust. When it does rain, this powder makes the water bead up, rather than being absorbed by the dry soil, a process not surprisingly known as 'skinning'. Water flows rapidly downhill over this skin, making a flood more dramatic; hence the term 'flash flood'.

The railway runs east to west crossing these sand-filled wadis, so wherever the rain falls, it washes out the railway somewhere. Always. To prepare for this eventuality, we had an airlift. Two or three C-130 'Hercules' flights a day from Khartoum, first bringing temporary housing units for the pilots and ground crew of the three Boeing 107 Vertols - the civilian version of the military Chinook - and then fuel and relief supplies. The fuel – JP1, an aviation grade kerosene – was discharged into 12,000-gallon bladders lying on the ground, like giant waterbeds.

FUEL BLADDER,

of the twelve-thousand-gallon variety

Once the infrastructure was in place, the airlift began. There were typically two with fuel for each food delivery flight, such was the thirst of the mighty double-rotor Vertols, as they shuttled back and forth to the limit of their ninety-mile range on their missions of mercy.

The cargo is carried in nets suspended under the helicopters, on the end of a 100+ foot cable. This is not just for ease of delivery and keeping the inside of the chopper clean - both of which it does - but the nets allow each bird to carry a heavier payload.

Here's how: by leaving the payload on the ground, the chopper can easily get airborne, warm up, and back up for a running start, or should I say a flying start, to literally snatch the cargo net off the ground at speed. When a helicopter is moving, it generates extra lift from the airflow over the rotor disc, which acts like the wing on an aircraft.

This extra lift is significant, and not available in hover. Getting from 'hover' to 'forward motion' for the extra lift is called 'transitioning,' and, needless to say, getting it right with a 5,000 pound payload in the 'bag' is an art in itself. Many times, the cargo net would bump and bounce along the ground as the overloaded machine, with engines redlined, struggled to get its cargo off the ground. Burning 50 lbs of fuel a minute, lift soon overcame load and the whole contraption lumbered into the sky.

The pilots and crew were Vietnam Vets to a man, inured to danger and fearful of boredom. With nowhere to go and nothing to do after the day's work was done, they soon discovered that my place was a good place to hang out, since it offered food, booze and a good card table for a poker game, which soon settled into a daily, or more precisely, a nightly routine. Let's face it, my place, primitive as it was, was a lot better than the alternative of their shoebox billets.

There was many an entertaining evening, often quite raucous, and between the poker and the war stories, we seldom had a dull moment. I remember one evening when the pilots were swapping crash stories from their collective experiences in action. In the final tally, they had been shot down a total of 14 times between them, for an average of 2.3 per pilot.

The most colorful was a Huey pilot describing a blade separation, when a chunk of one of his two rotor blades was shot away, throwing the rotor disc massively out of balance. The vibration shattered the glass of his gauges, and popped the instruments out of the panel, which then danced on the end of their wire connections like crazed marionettes. He and his crew, strapped into their seats and shaking so much they were unable to speak without bleating like sheep, enjoyed the half hour ride to safety in what he described as the world's most powerful 'vibrator.'

Both my outfit (Arkel-Talab) and Columbia Helicopters were under contract to Uncle Sam, and so were coordinated via the US Embassy in Khartoum. This had major advantages when it came to the logistics pipeline. Ours wasn't bad: an Air Safari Cessna 404 light twin, every two weeks, bringing our food, payroll cash, repair parts and such, and often with someone from Head Office or the Embassy to make sure that we had not 'gone native!'|

The pilot, along with any visitors, would overnight with us, departing first thing in the morning, back to Khartoum with receipts, reports and requisitions for the next trip. Departure was always at first light: the desert heats up fast after sunrise, triggering convective thermals that spawn the whirlwind dust devils which locals call Djinns.

As the day heats up, these hot air bubbles become more intense, rising higher by a thousand feet every hour, dragging with them dust and turbulence, and reducing visibility until the horizon disappears and any plane – and its hapless occupants – are bouncing around in a cocoon of brown haze, with even the ground obscured.

TYPICAL DARFUR VILLAGe
note the brown haze from thermals

Our little lifeline paled, however, in comparison to the giant Hercules cargo lifters, with their forty thousand pound payload capacity. The ability to hop on the Herc for a night on the town, and be back the next morning, was a real treat. Not that Khartoum was a hot spot . . . but compared to Nyala, it was like Paris to Peoria.

But the most fun of all were the helicopters.

Just before dawn, when air was the coolest, the choppers would fire up, hovering to bring the turbines up to operating temperature for max power. This usually took 10 to 15 minutes, perfect for hopping the quarter mile or so from the airfield to hover over my house, and for good measure – in case the sound of a twenty thousand pound, three thousand horsepower twin rotor helicopter hovering just over your head doesn't get your attention – they liked to drop the double steel hooks on the sling line on my corrugated iron roof.

Try sleeping in a Jamaican steel drum that someone is hitting with a sledge hammer . . . and then throw in a hangover!

Once aroused from slumber and my umbrage assuaged, I was usually able to get to the airport in time to hitch a ride to somewhere, Umm Dukun for example. Now Umm Dukun, which I think means Mother of Smoke in Arabic (God knows why), was a typical scruffy little village, little more than a few dozen round mud huts with thatched roofs. But it happened to be a few miles from the border of the Central African Republic (C.A.R.).

So what? You may well ask. Well, at the time the C.A.R had two claims to fame. First: as a former French colony, its first president Jean-Bedel Bokassa had had himself crowned Emperor Napoleon the Third, at the expense of over half the National Treasury; and second: during his 'reign,' the C.A.R was reputed to be the single largest buyer of French Champagne worldwide.

Apparently the Emperor liked to party.

Back then, Sudan lived under Sharia Law, where the penalty for alcoholic consumption was forty lashes outside the Mosque on Friday... despite which, proximity of a source of good booze was irresistible, and it seemed a waste to return an empty chopper with so much unused cargo capacity.

And to be honest, we had pretty good cover. (See story titled I am not Competent.)

So every two or three weeks, we made a delivery to Umm Dukun, to be greeted by the Mayor, Chief of Police and Major Pete, commander of the Army garrison.

WELCOME / RECEPTION COMMITTEE AT UMM DUKUN

After putting the cargo net on the ground and releasing the cable, the chopper would slide off to the side, land, and shut down. Signing off on the shipping documents took a few minutes, and then it was time for 'fatoor' (breakfast) so we retired to a table under a shade tree. Meals in this culture are social rituals and not to be rushed. No fast food in Darfur. . . and apart from our supplies, there was no food at all!

Over breakfast, we would discuss local issues, the condition of the farmers; whether the planting was going well, and always what news there was of the Jan-Jaweed. The Jan Jaweed are Bedouin, nomads, as opposed to agronomists or farmers.

There is a saying in the Middle East: "When the Oasis is strong, the Bedou will trade; when the Oasis is weak, the Bedou will raid." It defines the historical relationship between the nomad and the farmer, as any cinema buff who has seen The Seven Samurai or the Magnificent Seven will appreciate, and, by extrapolation, an understanding of Middle East politics.

The fertile crescent, as the coastal strip around the Eastern Mediterranean is historically known, is surrounded by desert where the nomadic Bedou, whose wealth is on the hoof, roam in search of food and water for their flocks.

At the time, the Jan-Jaweed had taken to raiding; whether from necessity or opportunity was a matter of debate. The reality was that Major Peter, the local garrison commander, was there to protect the oases; farming communities like Umm Dukun.

The problem was that the Jan-Jaweed was an unofficial instrument of the Central Government (with plausible deniability, of course) and was used to intimidate or punish any communities which sympathized or supported the Christian / Animist rebels in the South.

Well-armed and highly mobile, the Jan Jaweed were like ships of the desert, who "go where they please and strike where they please," as Colonel T. E. Lawrence of Arabia described his, or should I say Faisal's, Bedouin army. Major Peter's puny garrison was no match for the highly mobile Jan Jaweed, who plundered, raped and burned at will; explaining perhaps why the Major sought solace in whiskey.

Bonding took place over an unhurried breakfast, at which alcoholic beverages from the C.A.R. were available, sharia notwithstanding, Major Peter was particularly fond of whiskey, and if fuel were sufficient, a short aerial tour of the neighborhood helped cement the bonding. All of which was justified under the rubric of 'intelligence gathering.' It also established a back haul of fine consumables from the C.A.R., most of which were transported back to, or via, the Embassy in Khartoum . . . but a sufficient quantity was left in Nyala to satisfy the needs of the pilots and ground crew of the airlift, and, of course, the Arkel Talab field office personnel: both of them.

The trade was conducted via shortwave radio ahead of each trip. The exchange rate was established using the code 'tins' and 'boxes'. A box was a case of whiskey, wine, champagne or whatever; and a tin was a 55-gallon drum of kerosene: a rare and valuable local commodity used for lighting and cooking in the vast tracts of rural Africa, where there is no electricity.

Aviation Turbine Kerosene (ATK) is a very high-grade kerosene, and we had thousands of gallons of the stuff: need I say more? The 'tins' were delivered by truck ahead of the grain delivery, and the boxes from across the border were there to be loaded onto the chopper after 'fatoor'!

The return journey was usually routine. However, occasionally, when fuel permitted, we would divert to Jebel Marra, an extinct volcano not far off course. The mountain towers above the surrounding desert, topping out at 12,000 feet, and generating its own weather system by squeezing enough rain out of prevailing winds to create a lush green carpet of vegetation with streams, waterfalls and plunge pools . . . not to mention, plenty of wildlife.

Company picnic at Jebel Marra

It was baboons which provided the best entertainment at Jebel Marra, and there were a couple of pretty big troops in the forested areas around the base of the mountain. I profess no zoological expertise in animal behavior, but having employed one once for security (see: Reel in the Monkey), I am somewhat familiar with their behavior . . . not the least from an afternoon in Manyara, Tanzania with Alan Root, one of Africa's foremost wildlife photographer/filmmakers, observing a troop of baboons downwind and from a barely sensible distance.

Alan was born and raised in East Africa, was fluent in Swahili, and was totally comfortable in the bush, as the untamed wilderness is called. He had an almost spiritual connection with its animal occupants. Alan could imitate their sounds, and when recounting his experiences with wild animals, would adopt their facial expressions and even their physical postures; much to the entertainment of his audience.

He had adopted and raised several wild animal orphans, including a baby baboon, and so was eminently qualified to describe the behavior of the troop we were watching. He could identify the dominant male(s), their inner circle of desirable females, the pregnant and nursing females, all in concentric circles around the power center of the dominant males. The outer perimeter was patrolled by the young adolescent males, known as askaris (which means soldiers or policemen in both Swahili and Arabic). The askaris were disposable in the hierarchy, there to raise the alarm, and in the process give their lives for the troop in the event of an attack. Sound familiar?

On the ground where the food is, the askaris are vulnerable to big cats, especially leopards; as well as snakes, especially pythons. Interestingly, ophidiophobia (fear of snakes) is rampant among these, our simian relatives.

Safety lies in the trees, where their agility is supreme.

When foraging on the ground, a baboon troop immediately takes to the trees when an alarm is raised. Alan could imitate their various barks including the alarm call, and watching mass panic as the troop echoed the alarm and stampeded to the safety of the treetops was just too much fun. After a pause for the chatter to resume, a sort of 'what happened and who cried wolf' chatter, Alan would make some different noises, perhaps an all clear, and the troop would descend from the trees, chattering nervously, looking around for the source of the safety signal. The bold charged ahead to get to the feeding ground first, while the timid held back, proceeding cautiously.

After an hour or so of observation, we had identified the hierarchy of the troop and named them, ironically with the names of my superiors in the corporate hierarchy where I was embedded at the time.

With this in mind, chasing baboons in a giant helicopter, and watching the behavior of the troop – first threatened on the ground, and then in their treetop sanctuary – was hilarious and somehow eerily familiar to human behavior.

When the safety of the treetops is threatened by an enormous, whirling, growling bird-beast at eye level, the hapless baboons panicked, all order disintegrated and chaos ensued. Unsporting, perhaps, but highly entertaining.

How fragile is social order when self preservation triggers collective panic.

The 'Apocalypse Now' gang were with us for a few months, until some decision by the Sudanese Government offended the State Department enough to end the program abruptly and evacuate all personnel involved. With a few hours advance warning, I was able to notify my neighbor, who happened to be the Shell Oil dealer with fuel storage tanks.

The next morning, when the news was out, the unfenced airfield was invaded by hordes of looters, who made off with anything and everything left behind. By the time I arrived to survey the damage, which was substantial, there on the ground were two empty fuel bladders that had contained about 12,000 gallons of Aviation Turbine Kerosene (ATK) the day before.

My neighbor had a big grin on his face the next time I saw him, and was most helpful on several subsequent occasions, including when HRH Princess Anne arrived to visit Save The Children, UK, an NGO charity she sponsored.
SCF (UK) had several feeding centers that we supplied in and around Nyala. However, when the time came to depart, the turbo prop Vickers Viscount of the Royal Flight did not have enough fuel for the statutory range requirements, nor could SCF find any.

So, cap in hand, my friend Chris, field director of SCF, showed up begging for help. Fortunately, with the aid of my neighbor, we were able to find some aviation-grade fuel and enable the Princess to go home!

Neither my neighbor nor I got a knighthood – or even an invitation to tea at the Palace – but we did get an appreciative nod from Chris, the Darfur SCF field director.

Back in the U.S. some months later, the auditors caught up with me wanting to know what happened to the abandoned fuel. The looting provided plausible deniability for my poor recollection of the event.

VICKERS VISCOUNT OF THE ROYAL FLIGHT
Nyla 1985

HRH Princess Anne

under the Port Wing of the Royal Flight (Nyala Airport 1985); SCF Director Chris Eldridge in the dark blue shirt.

UMM DADA

Nyala, Darfur, Sudan 1985-86

It was hot. Very hot.

Day after day the sun beat down without mercy from a cloudless sky. The rail terminal a mile or so from the outskirts of Nyala was the end of the line. It was a single track, narrow-gauge rail line that followed the Nile from Khartoum to Wad Medani, and then wove its way across the desert via Kosti, An Ahud, Ed Dien and a dozen or so lesser stops before terminating in Nyala.

TRAIN STATION AT NYALA

Trains were infrequent and unscheduled and notification of their arrival was seldom more than 24 hours. Even then, the contents of the train were a surprise until the train arrived. This was because each of the dozen or so stops on the line had a shunt line, where rail cars were parked alowing east and west bound trains to pass and the local station chief to determined the priority of the rail cars that would constitute the 26-wagon maximum train that would proceed to the next station. Needless to say, rail cars could, and often did, get lost for weeks, months, or even longer, unless the 'Agent' was there to grease the bureaucratic wheels at each stop.

We - the company - had such a 'consultant,' who rode the train with a small suitcase containing sufficient lubricant to 'grease' the railway' from one end to the other. Hassan was a retired railway official. . . he knew the inner workings of the State Railroad system: its politics; its unions; its limitations which were legion; and its capabilities which were limited. He was worth his weight in gold getting relief supplies from Port Sudan to Nyala.

On arrival, our job was to assemble an army of porters, "checkers," and donkey carts to offload the train. The porters carried the 50 kg (110 lb) sacks from railcars to the waiting donkey carts, for transport to a nearby storage area. Checkers kept score at each stage of the logistics chain, recording the number of sacks offloaded onto each donkey cart and again from the cart to the storage yard. At the end of the day, porters and donkey cart drivers were paid based on the number of sacks delivered. This had to be accomplished in 24 hours to avoid huge demurrage charges, and, of course, had to be accomplished in daylight.

All the while, the unfenced rail yard teemed with hoards of indigents - mainly women and children - scavenging the spillage from damaged sacks, and, increasingly, using knives to enhance the spillage.

Trying to keep this ever-growing crowd at bay was an ever-growing problem. The Railway police, in their heavy parade-ground boots, were hopelessly outnumbered and out-maneuvered by the hordes of nimble barefoot and opportunistic urchins.

CHAOS AT THE RAIL YARD

What to do? We tried everything.

Meeting after meeting was held with the railway stationmaster, while endless cups of 'karkaday', a sweet red hibiscus tea, were served by Umm Karkaday, the tea lady who squatted on the platform outside the Stationmaster's office, endlessly brewing her product over a little charcoal brazier.

The stationmaster, Ahmed, was one of the sweetest human beings on the planet. The more we presented our problems, the more he wrung his hands and matched them with his own woes. At times I feared he would burst into tears, so overwhelming were his budget, personnel and administrative afflictions. But always at the moment of catharsis, he would shout "Jeeb chai!" (bring tea!) and Umm Karkaday would appear with more tea, and the tensions would subside.

When all official channels are dead ends (or cul-de-sacs), it is time to resort to the unconventional. In despair, I watched our deteriorating situation, passing the point of believing the problem could be solved without massive intervention beyond my pay grade. And then I noticed Umm Dada.

Among the women and children who swarmed around the rail cars, Umm Dada stood out. She was an imposing woman, and clearly a leader, but to be honest, it was not this quality that first caught my eye. She was a big woman with huge breasts that were scarcely concealed under her loose-fitting smock, and being a youngish man at the time I noticed such things.

Now the women in these parts of Africa do all the heavy lifting in life, and consequently have inordinately strong backs. Planting, hoeing, gathering or sweeping, African women spend hours every day of their lives, bent over with legs straight and backs horizontal, advancing slowly and rhythmically with the task at hand, and usually accompanied by a couple of toddlers and a baby slung on the back.

Umm Dada was no exception. Clutching a handful of reeds to make a whisk broom, she swept across the ground, herding every last grain of sorghum, her great untethered breasts swaying with sweep and step like mighty metronomes. Mesmerized, I watched in thrall her stately sweep and rolling gait as she worked her way across the spill zone, her competitors melting away like guilty children before the advance of big Mamma!

And then came the epiphany. Of course! She was the answer!

Through appropriate channels a meeting was convened, with suitable witnesses to ensure that no improprieties of Islamic protocol could be alleged or even rumored. Would she organize the ladies and keep them and their children out of the work area if she and her ladies were given exclusive access to the rail cars when we were done? It was inferred that we had no interest either in the spilled grain or the contents of damaged sacks.

Well! It did not take long for the implications to sink in. The next day Umm Dada and her chosen cohorts kept the crowd at bay with ruthless efficiency while we offloaded the train. Her dozen or so 'deputized' ladies with a score or more of fleet-footed urchins maintained a 'cordon sanitaire' around the work zone like a pride of lion defending their kill from a pack of marauding hyenas!

Railcar by railcar, as soon as the last sack was out, the signal was given, and Um Dada's ladies moved in to scour every last grain while the rest patiently waited their turn. Ahmed the stationmaster was delighted, and was soon taking credit for this role reversal, while his railway police, relieved of having to do actual work, resumed their habitual role of loafing around and sipping Umm Karkaday's tea, no doubt purchased with grain courtesy of Umm Dada. Sometimes the solution lies in the problem.

Breakfast, a broken Navaid, and a Saudi C130

Darfur, Sudan 1986

When the drought broke, as often happens, Mother Nature had the last laugh and delivered floods. Since the grain of the land runs north to south, and the railway ran east to west, it was predictable that the rains would produce flash floods and wash out the railway. Which is exactly what happened.

In anticipation of this inevitability, we had a contingency airlift planned. Before the age of GPS, finding a dirt strip in the middle of a desert – essentially devoid of landmarks – required a navigation beacon. A radio transmitter, called an ADF, was duly flown in, complete with a small portable generator.

The generator had to be attended for obvious reasons, and turned on and off at the appropriate times. There were typically three flights a day, of the big C130 Hercules Cargo Lifters, two with fuel for the three huge, thirsty twin-rotor Boeing Vertol Helicopters which shuttled relief supplies, and one with food.

Devoid of major landmarks, the western desert of Sudan is further obscured (navigation-wise) by heat-generated thermals raising clouds of dust aloft. At the time, there was a sort of precursor to GPS, called LORAN C, a global system of long-range radio beacons, three of which could provide a reasonable fix of one's location. Unfortunately only two of these beacons were detectable from Nyala, so the principle of triangulation was only marginally effective.

Eyeball navigation necessitated spotting the rail line and following it west, or spotting the Nyala-El Fasher dirt road and following it south, both of which converged on Nyala and neither of which was easily spotted through the desert haze. So the ADF beacon was provided, lest an expensive C130 get lost and forced to ditch in the desert.

Everything worked fine for a while. The local lad I assigned the duty of attending the little generator for the hour or so needed to guide an incoming flight, did his job until the aircraft had landed and then he shut down . . . until one day, an extremely irate C130 pilot confronted me with accusations only a little short of willful and potentially homicidal dereliction of duty!

It transpired that the ADF beacon had disappeared from his aircraft panel while he was still a few hundred miles out. I set off immediately to the place of the ADF and there it was: shut down, with no sign of the operator.

It didn't take long to locate the operator, who was bewildered by all the fuss. His reason for shutting down the generator was one word: fatoor (breakfast). For him, it was just about priorities. Sitting there watching a generator, versus fatoor, was an obvious 'no contest'.

In a culture where every meal is a social celebration, meals represent a communal gathering second only to religion. No grab-and-go fast food! This concept survives in America only in the annual Thanksgiving holiday feast. 'Fatoor' – from the Arabic root fatah – means "the opening," hence the "opening" meal of the day. Ironically, breakfast, – breaking the fast – means the same thing.

There were also other events which didn't go as planned.

The Saudis, eager to fulfill the 4th pillar* of their faith, alms to the poor, sent a C130 loaded with relief supplies to Genaina, on the Chad-

Sudan border where there was a large refugee camp. Its short dirt runway proved insufficient for the heavily-loaded bird, which almost came to a halt in time . . . but not quite.

WADI GENAINA, THE CHAD SUDAN BORDER
(airport top right)

At the end of the runway was a wadi, a river bed, about 10 feet or so below the grade of the runway. Shuddering to a halt – with props in full reverse thrust, brakes locked and skidding over the sand – the plane tipped gently over the edge, its nose down in the wadi and its tail high in the air like some great bird quenching its thirst.

Such events inevitably attract a crowd: some to gawk; and some, eager to help, bringing donkeys, camels, ropes and chains, and attaching them to all kinds of inappropriate and delicate parts of the aircraft, while the Saudi pilot tried desperately, but unsuccessfully to maintain control.

An attempt to bring a tank from the local army garrison failed, due to lack of fuel; and likewise, the bulldozer from the public works department was inoperable. And so the Saudi pilot abandoned the scene, leaving his plane in the tender care of a couple of gendarmes, and wandering off into town, his head down, and shoulders slumped in despair.

An hour or two later, he returned in a much happier mood, weaving slightly, it is true, having found a source of Araki, the local moonshine, prohibition of alcohol under Sharia notwithstanding.

The next day, a second Saudi C130 arrived, stopping short this time, with all the necessary equipment to save the day. An expensive enterprise!

THE SECOND SAUDI PLANE
arriving to the rescue of the first

*The 5 pillars of Islam are:

1. Profession of the Faith
2. Prayer (Salat)
3. Fasting (Sawm)
4. Alms to the Poor (Zakat)
5. Hadj (Pilgrimage to Mecca)

Chapter 4:

Uganda. 1986 - 1989

Tip-Toeing through the Minefield

Uganda 1986

Part 1

A few months after returning from Darfur, when I had caught up on sleep and regained the 30-odd pounds (avoirdupois) that I had left there, I remembered Fred Cuny's parting words as he handed me his business card: "If you ever get bored when you get home, give me a call."

Miraculously, I had come across his card while engaged in one of those periodic and futile attempts to get organized. So I did. Call him, that is.

The organization attempt would have to wait . . . "I wondered when you were going to call," he interrupted, while I was trying to remind him of our previous and only encounter, by now almost a year past in Nyala, Darfur. "How would you like to go to Uganda?"

And so it was without more ado, that a couple of days later I was on my way to Washington or Dallas or somewhere for a briefing, and the next day in Nairobi, Kenya for an overnight and more briefings, before heading to Entebbe, Uganda.

Uganda was a mess.

Idi Amin had recently been overthrown and Yawerri Musaveni had just graduated from insurgent to President and taken over the country. Needless to say, the country was in turmoil, especially the North, where the remnants of Idi Amin's goons and several other rebel groups were still running amok.

The rape and pillage by heavily-armed drugged and drunken thugs endemic in these civil conflicts produced a slew of orphans. Like stray dogs, these orphaned kids often hung around the gangsters' camps for food scraps and such, and increasingly were conditioned and used by the gangs.

Exploiting the natural human reluctance to shoot children, these evil monsters conditioned preteen children to commit heinous crimes against adults, even their own family members, thereby detaching them from any moral conscience, and used them as frontline troops who could infiltrate Police or Army lines by posing as victims.

Among the most demonic leaders of these children was a Voodoo priestess named Alice Lakwena, who we called "the Wicked Witch of Gulu." Her followers believed that as 'true believers' they were bulletproof, so that anyone who was shot was therefore not a true believer.

Another was Joseph Kony, founder of the 'Lord's Resistance Army' which also specialized in militarizing young children.

For the next two decades the various manifestations of these quasi-spiritual resistance movements survived every attempt to exterminate them, metastasizing like cancers into the neighboring states of Southern Sudan, the Central African Republic, Congo and Kenya. Adapted to Islam, one could be forgiven for wondering if the Boko Haram of Nigeria is yet another metastasis of the same cultural cancer.

There was little to choose between military and rebel forces in such conflict zones: both were undisciplined and used their weapons for extortion, interrogation and recruitment. Often, rival forces would hit the same community within hours or days of each other. Abandoned villages, or those with no young men, were the telltale clues of a 'hot' zone.

To add to the chaos, the eternal civil war in the south of neighboring

Sudan had hotted up. Our mission in the jargon of "UNILAND" was a 'Needs Assessment', an evaluation of the "push-pull" factors of the civil unrest, aka civil war. Cross-border population migrations are pushed to escape hardship, drought, famine, conflict, etc. ("push" factors); or drawn across borders by better prospects and resources: food, safety, employment, and yes, UN welfare ("pull" factors); often both.

Such refugee migrations always cause problems competing for scarce resources wherever they end up.

By estimating the numbers, and the likely destination of the refugees, relief agencies can prepare and respond in a timely fashion . . . or at least that's the idea. Migrations cause conflict. After all, when a bunch of destitute strangers show up in your village, and you have yams, and they are hungry, the milk of human kindness dries up pretty fast!

So our mission was to evaluate the situation on the ground, and provide estimates of where, when, and how many displaced people the UN should prepare for. It's a bit like trying to decide where to put the fire engine when the woods are really dry. Everybody knows there's likely to be a fire, but where, when, and what spark is going to start it, (and hence where best to park the fire engine), is a different kettle of fish.

Our employer was The UN High Commission for Refugees: UNHCR for short. We were a contribution from Uncle Sam, who frequently prefers to contribute "in kind" rather than in cash, for reasons that need not be enumerated.

With conditions perfect for population displacement, problems are inevitable as migrants compete with the indigenous for scarce resources, wherever they end up.

Nothing daunted Fred and I set off from our base, the White Rhino Hotel in Arua, one of those clusters of buildings that somehow qualifies as a town and gets on the map, in this case the top left or northwest corner of the map of Uganda, where it borders Zaire to the West and Sudan to the North.

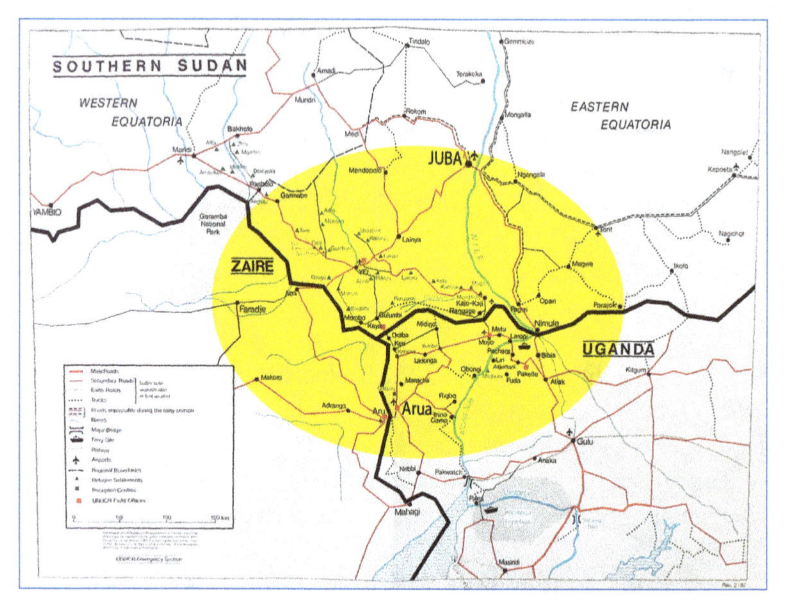

Map of Uganda, Zaire and Southern Sudan

We followed the dirt road north along the border with Zaire until it turned east, suggesting that we had reached the border of Sudan to the north. Our ride was a standard UN spec: Toyota Landcruiser, a diesel-powered brute equipped with oversized off-road tires, roof rack, two spare wheels, long-range tank (700 miles at the minimum) and frequently fitted with Kevlar plates to make the occupants feel better in mine-infested areas.

Unfortunately, it doesn't take long to figure out that it's the wheels that set off mines . . . and they're not protected . . . so the Kevlar bottom really doesn't do much for you, unless the vehicle is on its side, in which case it makes a pretty good bulletproof shield.

There were no borders in these regions – at least not on the ground – only on maps. And there are virtually no roads, just tracks, so when the trail turned to the east, we figured we were now skirting the border with Sudan, and started looking for a northbound track on our left. Soon enough, a likely candidate showed up and we turned north to follow a rutted track to its destination.

It was some time before we came across a person, or, for that matter, any sign of human habitation, until we came across a man with a machete, chopping wood. We stopped and after some chit-chat of the 'where are you from' and 'what's going on' variety, Fred asked him "Where does this road go, and who uses it?" to which the wood chopper replied: "No! De people do not use dis road any more because of de mines". It took a moment for this information to sink in.

Fred and I looked at each other, wide-eyed and in silence, along with great beads of perspiration, as the reality of our situation dawned on us.

We had been driving down a rutted track through thick bush for well over half an hour, or at least 10 miles, looking for a village or farm with locals who could help us collect the information that was the purpose of our mission. Our progress had been slowed by piles of rocks every few hundred yards. They were stacked in little pyramids right in the wheel tracks that we had to follow, forcing us to detour around them.

Needless to say we were irritated by this seemingly senseless interference and had concluded that someone was trying to slow down or damage vehicles, whether military or rebel we had no idea, perhaps as a prelude to ambush or robbery.

None of the possibilities was very encouraging so, invoking unprintable curses on the builders of the piles, we just kept going on the assumption that the track sooner or later would lead somewhere.

Eventually, we came across the man chopping wood who told us: "The people do not go down this road any more since they put in the mines." After letting this information sink in, he continued in his broken English "but some of the people were hiding in the bush watching them, and after they went away, they put stones on the places where they put the mines"... ! Hearing this, we quickly revoked the curses we had muttered on the stone pile builders!

We had set out that morning from Arua in north Uganda, and were well inside southern Sudan. The border between the two countries is well defined on maps, but no trace can be found on the ground. The long-festering civil war in the south of Sudan was hotting up, and Northern Uganda was a lawless region where the remnants of Idi Amin's goons – the Lord's Resistance Army, or LRA* about which more later, gangs of armed poachers, and renegade army units from both Sudan and Uganda – roamed around poaching and pillaging, more or less at will.

PASTORAL SCENE IN NORTHERN UGANDA
In the early post Idi Amin era of the late 1980s.

Our problem now was how to get back without getting blown up. There was no alternate route and the bush was way too dense to try to force a new track, so there was nothing left to do but return the way we had come.

It was dry, but our tire marks were still just visible, so we would have to follow exactly in our tracks and hope for the best. That meant, however, that one of us would have to sit on the bonnet (British for the hood) and the other would drive.

Fred pulled rank on me and chose to drive, so I sat over the front wheel consoling myself that at least I would be right over the blast for a quick clean kill, whereas Fred, slightly protected in the cab, would face a more lingering fate. With unbelievable focus, I stared down at the faint tire tracks waving my hands to signal minute corrections left or right, as we crept back down the road following EXACTLY in the tracks we had made going the other way.

And that is how, eventually, we made it back and earned a well-deserved beer at the White Rhino hotel in Arua.

The White Rhino had once been an oasis from which wildlife safaris into the nearby Kidepo National park were launched. Needless to say, the tourists had disappeared, as had the game, mowed down for ivory, skins or sheer drunken entertainment by rebels and both Ugandan and Sudanese Army units.

We made several more trips up and down the border gathering as much info from the field as we could, entering, on one occasion, an area infested with tsetse flies. The tsetse fly is a large aggressive insect like a horse fly, with a bite that really gets one's attention. Worse still, it carries the dreaded trypanosomiasis, better known as African sleeping sickness, which turns its victims into walking Zombies.

Prior to independence the situation was under control, at least to the extent that the tsetse fly areas were shrinking and cattle farming areas were expanding, largely, it was true, from DDT.
After Independence, the programs were defunded and DDT banned, following Rachel Carson's viral best seller: Silent Spring. So the tsetse fly territory advanced, rendering vast areas of potential farmland unusable.

Driving through these areas with four or five people on board – windows closed to keep out the swarms of tsetses and with no a/c – everyone was equipped with a rolled-up newspaper to attack the odd tsetse fly which inevitably came in through the air intakes. Imagine, if you can, a flailing and writhing group of sweating passengers, desperately trying to squash these insects before they could bite someone. It was as hilarious as it was deadly serious.

Everywhere we went the extent of the looting was obvious. Corrugated iron roofing and copper wire were hot items, so metal roofing was rare as were copper wires for both power and phone lines. Even my room in the White Rhino Hotel, our base, had half the roof missing, exposing the African sky, while the bed was situated under the remaining portion of the roof. The room rate, however, remained $100 US.

Strangely, two institutions seemed to escape the looting: monasteries and breweries, something to do with spiritual refreshment, perhaps. One of the monasteries we visited was composed of an Order who worked the land, farming if you will. The monks, at least those whom I met, were hard working, chain smoking, no-nonsense types who infected the locals with their work ethic as much as their religious devotion.

The sanctity of the breweries was understandable but the respect commanded over rampaging savages by these down-to-earth monks was something different, and cause for reflection. Connected to a higher cause, they were not easily intimidated. On the contrary, they were intimidating, but their close association with the locals made them an excellent source of information.

It was a short mission, just a few weeks. We managed to scavenge some information for the report, but the really relevant information was the plans by the military or the rebel forces . . . and to find that out I had to go to Madison, Wisconsin on a subsequent and totally unrelated mission.

PART 2:
LOOK AT THOSE 'BLACKIES' OVER THERE!

A week after the Uganda mission, and before I had been able to finish my report, I was in Madison, Wisconsin, attending a semi-annual UN Disaster Management seminar for field operatives, at which I taught the logistics component.

Disaster logistics is based on two assumptions:
One) That the supply chain is broken so nothing works; and
Two) The rule of 3s.

The rule of 3s is the determinant of priorities. They are: 3 minutes, 3 days and 3 weeks. Let me explain. It's about air, water and food.

1) **Three minutes without air**, you're dead.
Pyroclastic flows, volcanoes, drowning from tsunamis etc.
Sorry, not much we can do about that post facto.
We're working on preemptive solutions but not there yet.
2) **Three days without water**, you're dead.
Potable water we can do, so don't drink bad water and start a cholera epidemic.
We can do clean drinking water in under three days.
3) **Three weeks without food**, you're dead.
No problem there.
Plenty of time, you may lose a little weight, but so what?

It was shortly before Christmas and very cold, as Madisonians are well aware. I was in a K-Mart store with a colleague from Finland, an old

Africa hand who suddenly said in his thick Finnish accent: "Look at those 'Blackies' over there."

Not quite sure to what he was inferring, I followed his gaze to a black family: father, mother and two young girls, who were perusing a carousel of kids' dresses. "I'll bet you they're from Wau" he said.

Wau is a small town in southern Sudan. So to settle the issue, we went over and introduced ourselves. Turned out they were from Yei, a small town in southern Sudan about 60 miles from Wau, or the other way 'round, but no matters, you get the point!

Long story short: the father's roommate in college had been John Garang, leader of the Sudanese People's Liberation Army, or the SPLA, the rebel organization seeking secession from Sudan, now a 'fait accompli'. He was studying International Law at Madison University for a posting to the OAU, the Organisation of African Unity, and was very close to the SPLA and its leadership.

So we met later at a nice pub, after the kids had been put to bed, had a great evening and got all the answers that we had been unable to ascertain on the ground. Who would have guessed? I was able to furnish a report with information that we could not collect from the ground, and thereby immeasurably enhanced my reputation for intelligence.

I never revealed the source of my information, of course, lest my insight be attributed to blind luck rather than intuitive genius.

Moral: He who knows not, and knows that he knows not, can be taught. Teach him.

The Budongo Forest

Uganda 1989

"I want to get out now."

These words, from the young woman riding with us in the back of our UN Land Crusher, were not what we wanted to hear. To my right, in the seat beside me, the whites of David's eyes shone like moons in the twilight gloom, from the dark ebony oval of his face.

Across the muddy trail in front of us were a couple dozen casually-dressed young men and women sporting AKs, RPGs, crossed ammo belts and other indications that they were probably not peacefully motivated.

Very early that same morning, David, my Ugandan interpreter, guide / driver / whatever, and soon to become – through that strange bonding that comes only from shared adversity – my bosom buddy, had set out from the Hotel Masindi in, of all places, Masindi. The 'Hotel' was in fact an old "Rest House" from colonial times, where the District Commissioner would rest overnight when touring his district.

It was a simple ground-level block house with two or three rooms and a mess hall or dining room. There was a caretaker or two who got to live in the house between visits, and cook and clean for the official visitors. Our mission was to explore alternate routes for the supply of emergency stuff to the beleaguered garrison town of Juba in Southern Sudan, at the time cut off from the world by the rebel forces of John Garang's Sudanese People's Liberation Army (SPLA for short) and being sustained by a very expensive airlift from Entebbe, Uganda.

We had covered a lot of ground that day. The monsoon rains were over, and the rutted routes we explored had baked hard from the last truck to make it through the mud before the monsoon rains made them impassable. Our off-road UN Land Cruiser was no match for the deep ruts in the muddy road along the shore of lake Albert, when late in the day it was time to head home.

According to our old Ordnance Survey map from Colonial times, there was a track through the BuDongo Forest which looked a lot shorter than back-tracking the roundabout route by which we had come. So we took our chances and headed into the Forest.
The Budongo is a tropical rain forest. Dense foliage canopy shuts out the sun except for the occasional sunbeam that sneaks through, to shed just a little light on its dark damp interior. The rain-soaked damp leaf mold floor feeds the fungi that break down the organic detritus in an endless cycle of growth and decay, while the fetid air, laden with moisture, provides a never-ending source of condensate that drips in heavy dew drops from the canopy above.

Once in the forest, the late afternoon light turned swiftly into a twilight gloom. The terrain was uneven, the trail seeming to follow a hillside with alternating ridges and valleys. As we progressed deeper, the ridges and valleys got increasingly steep, and as the track was muddy we dared not stop . . . except on the ridges, for fear of getting stuck in the valleys.

After half an hour or so, we had to turn on the headlights, and I remember trying to follow the road with only occasional glimpses between the headlights pointing up at the canopy and then down to the next ravine, as we bucked our way along the switchback trail. Several times we came over a crest and looked down into a ravine ahead with a rickety bridge at the bottom.

Such occasions called for a quick go/no go decision, since the only chance of getting up the other side was to build up enough speed on the down run to get up the slippery slope on the other side, and just hope that the wooden bridge didn't collapse. Several times the trail split, and whenever this happened, we chose the uphill option on the assumption that it would be less muddy.

At first, we passed the odd pedestrian carrying a load of something on their head: yams; a five gallon jerry can of water; or a branch of the ever popular matoki (giant green banana-like plantains). As we plunged deeper into the forest and the darkness descended, the number of pedestrians became fewer and further between, and we began to have reservations about the wisdom of our choice of routes, and so decided to stop the next one and ask directions.

Not long thereafter we passed a young woman. She spoke good English and offered to direct us to Masindi if we would give her a ride. She climbed into the back seat and directed us with authority: left here, turn right, etc. A couple of times, she took us downslope, which worried me a bit, but by then we had no option but to trust her. After about twenty minutes, we made an abrupt turn off the track and into a clearing, when the headlights lit up the dozen or so well armed bandits.

"I want to get out now," said our passenger. It was pretty obvious that our encounter was no accident, and after a brief assessment of our options – which were slim to none – I reached back and opened her door. She climbed out and walked straight over to a heavily armed lad in the center of the group, who bore all the characteristics of the alpha male of the gang.

After a few minutes, she, the alpha male and a couple of backups walked over, their weapons 'ported' or carried (as they say in military jargon) . . . a much more comfortable situation than having them pointed at you!

"Jambo! Habarree!" I called out the open window, trying to convey the impression that this was a perfectly normal situation. "Habarree Massouri!" came the Swahili reply to my greeting. So far so good!

But we still had no idea where we were, let alone how to get out of there. It soon became apparent that our passenger was the alpha male's girl and had convinced him that we were neither a threat nor good ransom material . . . or perhaps that one good turn deserves another. We'll never know.

Anyway, somehow we managed to preserve what I call the 'dumb foreigner' impression, and convince them that we were there to help the people, and were not an instrument of the government, or any of the rebel entities.

I had little trouble playing the ' dumb foreigner' role, which comes naturally to me, in addition to being well-rehearsed. David, being obviously local, was another matter, however. It took quite a while to convince them that he was a UN interpreter from Entebbe, and not a spy, informer or agent of the CIA.

Somehow we prevailed, and one of the lads was assigned to guide us back to the main trail, climbing into the back seat with his AK47. He rode with us for a mile or so, until we reached the main road, and then after giving us directions to Masindi, he set off on foot in the dark to rejoin his gang.

By the time we got back to the hotel, it was late; maybe ten or eleven pm . . . long after dark. It was clear that there was a level of anxiety among the hotel personnel (understandably since we were the only guests, and guests were few and far between). They also had the responsibility to prepare dinner, and in our absence did not know when to start cooking!

It was only when we explained, by way of apology, that we were late because we had come through the Budongo and got lost, that the alarm really set in.

At first they did not believe us. "Nobody goes into the B'Dongo and comes out alive!" they exclaimed. "It is full of bandits!" Then they told us that only a few weeks prior, the Army had made a sweep and captured a number of the rebels who were publicly hanged . . . in front of their families, who were required to attend. So we were told.

I am glad that we did not know this at the time.

Had we known, I doubt we would have had the 'sang froid' to bluff our way out of the Ambush.

Photo of Hotel Masindi

Down the Drain!

Uganda 1989

The decades-old civil war in Southern Sudan was on again. The southern capital Juba, encircled and besieged by rebel forces, was being sustained by a very expensive airlift out of Entebbe, Uganda, its neighbor to the south.

Supplies of food, medicine, ammunition and essentials such as a tennis racquet for a French doctor from MSF* (on one trip) were being flown in on Lockheed's giant C130 Hercules transports operated by Southern Air Transport, a Miami-based enterprise that emerged Phoenix-like from the ashes of Air America, shortly after its disappearance in a burst of publicity.

The Cold War was still hot, although under Gorbachov it was showing signs of cooling down a bit. Nonetheless, the main ingredients were still in place. Egypt and its southern neighbor Sudan were on our side, and Ethiopia was on their side. So there were plenty of surrogate wars to keep Disaster Relief folks busy.

The way it worked in the Horn of Africa was like this: the rebellions in the Northern Ethiopian provinces of Tigray and Eritrea, the latter a war of secession, were fighting the Soviet-backed Mengistu Regime, known as the Derg, supplied from bases in The Sudan, to which the Sudanese government turned a couple of blind eyes.

At the same time the Rebellion in Southern Sudan, also a war of secession from the Moslem government in Khartoum, was supplied from bases in Southern Ethiopia that the Ethiopian Government professed to know nothing about. It was a sort of round robin tit-for-tat game of 'you kick my butt, I'll kick yours'!

In both cases the rebels had control of the ground, especially by night when air cover was not available. Consequently government forces were cut off and besieged in garrison towns; Juba in Sudan, and Shirraro and Asmara in Tigray and Eritrea respectively.

Juba lies on the Nile, the 4,200 mile artery that, for millennia, has carried its watery lifeblood through the Sahara to nourish the Ancient Egyptian civilization that at least in major part spawned Greece, Rome and Europe.

Incredibly, for centuries, the source of the Nile was one of the World's great mysteries until, in the 1850s, Burton and Speke traced its source to the outflow from lake Victoria.

Ptolemy's second Century AD map of Terra Cognita, (The Known World) was a better representation of the heart of 'Darkest Africa' than any map for the next 16 centuries.

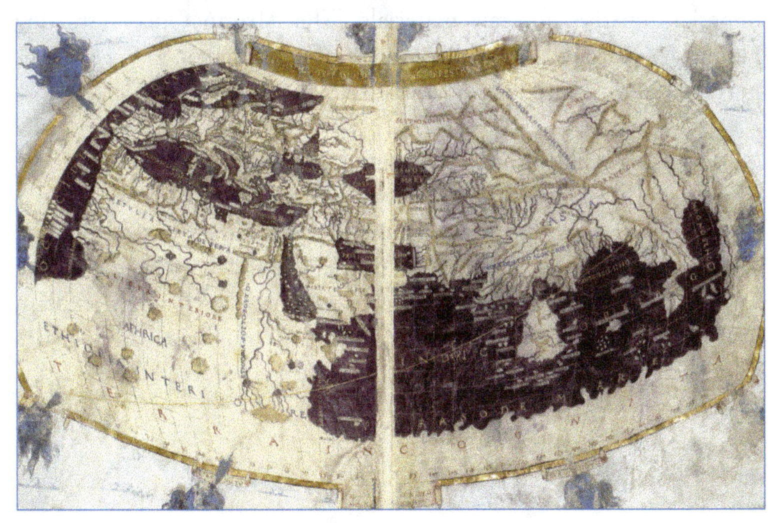

PTOLEMY'S 2ND CENTURY MAP
of 'Terra Cognita'

The reason is the Jonglei, a vast swamp that waxes and wanes with the seasonal monsoon rains that water the mile-high East African plateau and feed the Congo, Zambezi and the Nile: all but one of Africa's four great rivers. After Juba, the Nile literally disappears into the Jonglei, emptying into a network of swamps and streams not unlike the 'river of grass' in South Florida better known as the Everglades.

It was this great swamp that kept the source of the Nile a secret for so many centuries when rivers were the highways of travel and exploration through the jungle.

Fast forward to the late 20th Century. The shortest and most practical supply route to Juba was through Uganda, Sudan's southern neighbor, and in the absence of an over-land option, by a very expensive airlift out of Entebbe.

Our mission was to find an alternate overland supply route with a better Co$t-per-ton ratio, in the process of which we made some surprising discoveries. The region in the early 1900s had been a busy river trading region when the Industrializing European nations (Britain, France, Belgium and Germany) were busy gobbling up Africa to satisfy their Imperial appetites for raw materials. 'The African Queen,' starring Humpty Gocart and Katherine Hepburn, was based on C.S. Forester's novel, set in this time and place.

Great rusting hulks of once-substantial vessels – shallow draft paddle steamers for the river work and deep draft vessels for the tempestuous lakes – littered the shallows around the one-time port of Butiaba on Lake Albert. Strangely, none of the UN folk to whom we were contracted were aware of the region's history, which might explain why my suggestion that the meandering Nile might be the logistics solution, and not the problem, were met with bewildered stares.

LAKE TRANSPORT SHIP

PADDLE STEAMER

In between repairing broken bridges, assembling a ferry across the Nile, reconnoitering land routes and getting lost in the BuDongo forest, I was able to hitch a few rides on the Hercs working the Entebbe-to-Juba airlift, which was great fun.

Oops! Wrong Runway

Now the runway at Entebbe was the morning gathering place for various large carrion birds – vultures, marabou storks and such – waiting for the sun to heat up the runway to generate the thermals that lifted them effortlessly aloft, where they could spend the day cruising around on rising air columns and looking for dead things for breakfast.

Vultures have very low maneuverability and are really messy when struck by aircraft propellers or sucked into turbine air intakes. Consequently, a squad of Ugandans on bicycles equipped with horns, metal trash can lids and a variety of wind and percussion noise makers would ride up and down the runway, making a hell of a racket to scatter the flocks and clear the decks, so to speak.

Then we would take off for Juba.

Taking off without a bird strike was the first hurdle; the next one was

SAM2s. The SAM2 is a Russian Surface-to-Air Missile with, we were told, an operational ceiling of 16,000 feet, so we made sure to cross the border well above 16K altitude.

Supposedly, there was a five mile 'safety' radius around the Juba airport, but big old cargo carriers don't do tight turns easily, so the descent was called going Down the Drain involving 60 degrees of bank.

At 60 degrees of bank, you are pulling two Gs. Everything is twice as heavy, so it's a good thing to be sitting down, to avoid passing out.

Once on the ground, there was no hanging around: it is the most vulnerable time. Any hostiles in the area are now awake and activating their artillery or mortars, which may already be ranged. If the zone is 'hot' (stuff is incoming), the cargo is offloaded on the run, and takeoff follows immediately after the last item is out of the plane.

Safely back over the border, the fun began. Dropping down to 50 feet over the Nile, the barn door at the back was lowered (as if for an airdrop) and we, the cargo crew, would put on the monkey rigs (safety harness) and sit on the ramp, legs dangling over the edge, to watch the show.

At around 300 knots, you couldn't hear the plane coming until it arrived overhead in an explosion of sound. Watching hippos wallowing peacefully in the river, then exploding out of the water like polaris missiles was spectacular entertainment.

And then there was the Ugandan Army camp at Pakwatch.

Pakwach is where the Nile converges with the outflow from Lake Albert.** On the west side of the river was a Ugandan Army camp. Upstream a mile or so is Para Lodge, where we were installing a ferry capable of carrying a couple of ten-ton trucks across the mile-wide Nile.

Unfortunately, the crossing landed on the east bank of the Nile: a quadrant of complete lawlessness, especially after dark. Convoys through this area required military escorts, unless you had a high risk temperature or were fleet of foot.

When operating in a single vehicle, one was especially vulnerable. The trick, when confronted with the sharp end of an Ak47, was to grin and engage in dialogue as though everything was totally normal, with greetings in Swahili, like 'jambo habari' and then keeping the initiative with rapid fire questions like, "What are you fighting for?" to which the reply was usually: "We are fighting for free-dumb to the last drop of blood ' in that delightful staccato Rap-like accent.

Maintaining command of the conversation with an offering of irrelevant value to the cause, it was usually possible to escape without being pursued by a hail of bullets, while they puzzled over the paper notes they had been given in pursuit of the cause.

I made several excursions in the lawless quadrant, usually accompanied by an armed guard. My favorite guy was a short, stocky fireplug Ugandan with a double chevron on his jacket, that suggested that he was, at one time at least, a legitimate army corporal. He had one of those round oval coal-black faces typical of the Acholi, Idi Amin's Northern Uganda tribe. He was always grinning, which was nice, except that he had those dark unreadable eyes that made you wonder whether he liked you; thought you might be fun to torture; or maybe even good to eat! So I addressed him as "General," which he seemed to like, and we got along fine. I remember an occasion when I had been exploring something in the lawless quarter with the 'General' and we arrived late in the afternoon at the ferry crossing. The ferry was nowhere to be seen. A mile away, across the river, we could see the ferryman tucking his boat away for the night and studiously ignoring us. Nearby, was a ten-foot dugout canoe which the 'General' suggested as an option.

Having seen the twenty-foot Nile crocs in the region, I was not ready to challenge them in a ten-foot dugout canoe paddled by the butt of an AK47. So when no amount of hollering got the attention of the folks across the Nile, the General loosed off a burst from his AK. When nothing happened, he loosed off another burst with a higher elevation.

Maybe the second volley landed in the cluster of huts, who knows, but folks started running around waving at us, and in no time the ferry was on its way to rescue us. Disembarking a few minutes later on the safe side, I thanked all and saluted the 'General' who just grinned . . . whether because he liked me or wanted to eat me, I'm still not sure.

But back to the airlift, buzzing the Pakwatch Military Camp at 50 feet was too much fun. Pandemonium! Guys bursting out of bivouacs with one boot on, hopping around half dressed, trying to get a shot off at us. Occasionally, we would see a few muzzle flashes but by then we were well out of range, and besides, the muzzle velocity of an AK was insufficient to catch a receding C130 that wasn't slowing down.

Back at Entebbe, we tried to behave, and did, for the most part, except for one occasion when the Herc was taxiing to the hangar. The ground marshal was waving his ping pong paddles to indicate that he wanted us to park somewhere other than in the hangar the pilot had chosen. Refusing to follow instructions, our pilot edged straight ahead toward the open hangar door, while the steadfast ground controller held his ground like a student in Tiananmen Square.

When the gap between man and machine was down to a few yards and no one was backing down, the pilot hit reverse thrust and opened the throttles wide. The blast blew the ground controller fifty yards back into the hangar followed by the Herc, which wound up in the hangar of its original choice.

*MSF: Medecins Sans Frontiers, the French Doctors Without Borders.

MAP OF THE UPPER NILE

Historical footnote:

** During the reign of Queen Victoria (1837-1901), at the height of the British Empire, Burton and Speke finally established the source of the Nile, the fountain of European civilization, as the outflow of a great lake which still, miraculously bears the Queen's name: Lake Victoria. Downstream a bit, the now-called White Nile flows into a rift lake called Lake Albert, after Queen Victoria's beloved consort Prince Albert. (Back in the day women who married Kings became Queens, but men who married Queens became Prince Consorts . . . go figure ?!)

The Bamboo TransAll C160, A Jungle Genius

Northern Uganda 1989

The civil war in Southern Sudan, coupled with the collapse of law and order in the north of Uganda, had triggered large migrations as a helpless populace – seeking safe haven – fled the conflict areas.

Inevitably, such migrations cause problems, as destitute refugees flood into towns and villages in uncontested areas that are unable to accommodate them. There are no national borders in these places: such conveniences exist only on maps.

An abundance of NGOs and International agencies, such as UNHCR, follow these events with humanitarian relief, setting up temporary camps with shelter, food, water and medicine . . . all of which requires a lot of logistics: moving stuff from where it is to where it was needed, or, in some cases, moving the people from where they are to where the stuff is. Since logistics was my dodge, as they say, I got roped into a number of such situations.

One particular incident stands out, in what was otherwise a fairly routine operation.

We were airlifting several hundred displaced people from an area without amenities to a 'refugee' camp. Now the word 'refugee,' in its original meaning, describes people seeking 'refuge,' i.e. safety, presumably from somewhere that is not safe! However, in the legal jargon of the International Community, one has to cross an international border to qualify as a refugee. This has everything to do with 'whose responsibility are they' and nothing to do with their safety! Hence my use of the term 'refugee' refers to people seeking refuge, regardless of whether they had crossed a border or were internally displaced: two completely different legal descriptions of the same condition.

But I digress.

This particular airlift was carried out by a NATO military aircraft called a C160 TransAll. Similar but slightly smaller than the Lockheed Hercules, the TransAll is a big-bellied high-wing beast powered by twin turbo-prop engines that can be feathered into reverse thrust, giving it remarkable short field capability on unpaved landing areas where brakes are ineffective.

The huge rear cargo door opens to become a ramp, making loading and off-loading a full charge of passengers, with all their carry-ons, a matter of minutes . . . the envy of commercial airlines! With such a quick turnaround and a short round trip, we were able to cycle the plane five or so times a day, including a couple of diversions to a third location for refueling. The whole operation was completed in a couple or three days, as I recall.

Needless to say, the airlift caused a lot of excitement locally, attracting quite a crowd, especially kids, from surrounding villages, some of them several miles away. The area was dirt poor; there was barely a shoe or a shirt to be seen, especially on the kids. Fortunately, there were no facilities requiring shoes and shirts for service!

The entry level footwear was the flip-flop, which some of the adults sported, but never the kids. What the kids did with flip flops was make toys, there being no Toys R Us store for thousands of miles. Their favorite toy was a truck, presumably because the ambition of every young Ugandan lad was to be a truck driver when he grew up. Toy trucks were made out of whatever was available, typically sheet metal from old tin cans hammered into shape. Wheels were cut out of flip flops, using a tin can in the manner of a cookie cutter.

The toy truck was on the end of a stick which allowed the owner/driver to rush about 'driving' his truck, while making appropriate engine and klaxon noises! Some of the toys were well enough crafted to be recognizable as a Ford, Bedford or a French Berliet, all common off-road trucks of the time and place.

A few weeks after the airlift, I returned to the location for a follow-up visit, and was astonished to see a young boy (he couldn't have been much older than 11 or 12), playing with a model of a C160 TransAll. He was running around, holding his C160 TransAll replica high, making all the right engine noises, banking it perfectly for the turns. He and his homemade toy airplane were clearly the stars of the playground.

The model was perfect in scale and exquisite in detail. It had a working cargo door, the big 'barn door' at the rear of the fuselage, that opens downwards to form the loading ramp on the ground or the launch platform for air drops.

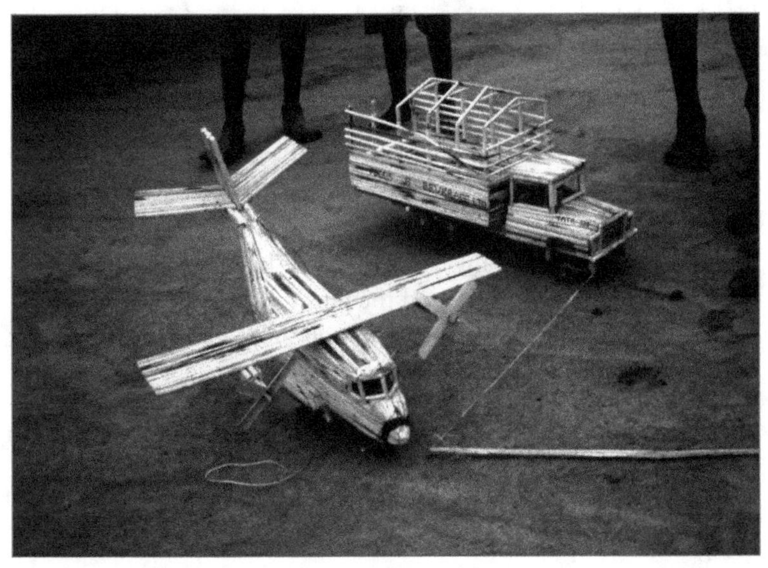

FRENCH BERLIET TRUCK AND C160 TRANSALL
homemade models

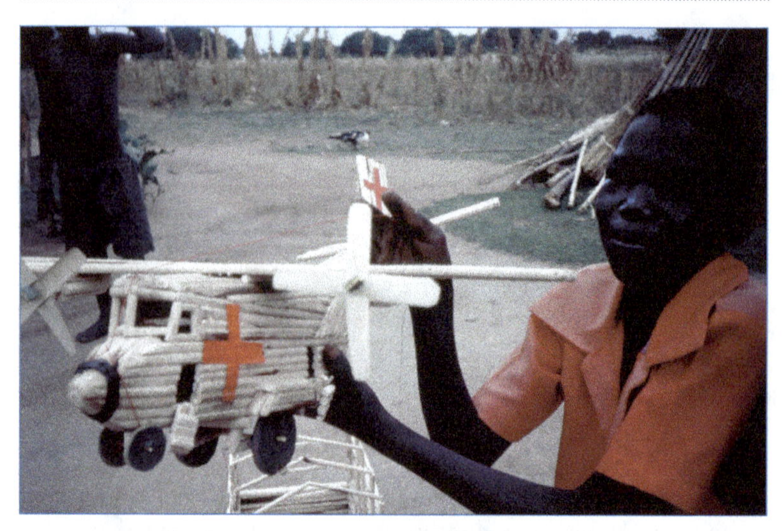

CLOSE UP OF THE C160 TRANSALL

Forward under the port wing was the crew door, which opened just like the real thing into a stairway. But beyond the precision of its made-from-memory form, it was the toy plane's construction technique that amazed my engineering mind, for the sheer genius of its fabrication from the limited materials available.

The entire model was built of bamboo using the thin tips of the stalks where the outer shell is hard and the center is filled with a soft pith. The pith stabilizes the pins used to tie the bamboo stems together, like planks to make surfaces, and the pins were made of slivers of the bamboo's hard outer casing . . . a one-stop shop! His mother's sewing needle for a drill, and perhaps his father's vise or clamps (he did allow that his father had helped him) were all he had had to work with.

On another occasion, also in Northern Uganda, I came across a group of young lads, obviously a band for they each had homemade multi-stringed banjo type instruments. The instruments were made out of a gourd, over which was stretched a goat skin. The bridge was a curved stick, and the strings were polyester chords from old tires tensioned by twist pegs in the bridge. When the ensemble started to play, the sound and rhythms could have been a Jamaican steel band, with that characteristic metallic plinky-plink sound. It was amazing.

I often reflect on that experience and wonder what will happen to that particular strain of genius: the genius of improvisation, born of scarcity, weaned on observation and recall, without the aide memoire of note, photo or iPhone . . . when everything is available off the GoogleWicki Internet shelf . . . ?

Ugandan boys band on the road
with homemade musical instruments

TPLF Elementary School Work Book

CHAPTER 5:

Ethiopia. 1990

A Tale of Two Nuns: Part I

Sudan/Ethiopia 1990

"Margaret! This is the man Jesus has sent to take us into Ethiopia."

These words, in a lovely Irish accent, greeted me as I walked into the dingey Rest House in Kassela, a dusty little Sudanese town on the Ethiopian border.

I had just arrived from Khartoum, after a battering, day-long drive down one of Sudan's 'better' roads . . . meaning that it was mostly paved, except for the potholes and other parts that were not.

The specially equipped Toyota Landcruiser I was driving was designed for rugged off-road travel and not passenger comfort, so my hindquarters had already taken quite a beating that day. In addition, I was carrying a cargo that made me distinctly nervous. The last thing I wanted was the added responsibility of two middle-aged Irish nuns on their first time in Africa – and in a war zone to boot – should I need to save my own precious hide.

Sisters Brigitte and Margaret had already been closeted in the Rest House for a week or more, waiting for the next episode of their mission to serve in a local health clinic somewhere in war-torn Northern Ethiopia. By the time I arrived, they were getting pretty desperate.

The Rest House attendant, called a chokidar, lived in a hut inside the walled compound, and served as guard and cook but spoke no more

English than the nuns spoke Arabic. He spent his time squatting over a charcoal brazier, boiling water to make tea and preparing the three daily meals provided for the guests.

The meals consisted of various combinations of eggs, the local flat pancake bread called Aseeda, and assorted goat stews; and as often happens in these situations, Margaret was suffering from an affliction known in the expat community as "The Mahdi's revenge."

This condition involved spending an inordinate amount of time in the latrine.

She also looked really miserable despite a slightly moustached stiff upper lip!

Now for those who slept through their history class, in 1885, the Mahdi, or 'Chosen One' led the revolt that, after a ten-month siege, annihilated the Anglo-Egyptian garrison under General 'Chinese' Gordon of Khartoum. Gordon's severed head was mounted on a pole and paraded around Khartoum to the delight of the faithful. Mohamed Ahmed, the self-professed Mahdi, met his 'Waterloo' over a decade later on September 2, 1898, at the battle of Omdurman – now a suburb of Khartoum – with the arrival of the relief force under Lord Kitchener... better late than never!

With Kitchener at Omdurman was a young cavalry officer who would profoundly influence history in the next century. His name was Winston Churchill. But I digress.

Brigitte's and Margaret's only other contact with the outside world was a secretive official of the Tigrayan People's Liberation Front (TPLF for short), who spoke a little English, and dropped in at random every couple of days to check on them, and assure them (without any mention of when, how, or where) that everything was going according to plan. All of which helps to explain why my arrival activated such a flutter of excitement.

Not that there was anything to do or see in Kassala. Dry and dusty, the main street was the road to Port Sudan on the Red Sea, some 100 clicks to the east; and Gedaref, the grain centre of Sudan, some 50 clicks to the west, where much of the country's staple sorghum was grown when rain was sufficient.

Most of the secretive cartel of grain merchants were headquartered in Gedaref, from whose secret buried silos seed corn was stashed and sold to desperate farmers at exorbitant rates during the not infrequent droughts. Typically, the farmer was obligated to hand over much of his crop should he be lucky enough to have one.

Motor repair shops; storage depots; stray goats and dogs; and – as the day heated up – dust devils, those heat triggered whirlwinds called djinis by the locals, that stirred up the dust and the plastic bags, that littered the streets; was about all there was in Kassala, except for the oppressive blanket of secrecy.

Everything was hush hush, nobody answered questions, nobody knew anything, nobody spoke to strangers, or even made eye contact, because Kassala was the main staging area supplying the TPLF in their decade-long, brutal struggle with the Soviet-backed Ethiopian regime of Haile Miriam Mengistu.

I was the only guest checked into the 'hotel' just down the street, apart from my Ethiopian guide who seldom spent more than half an hour there, before disappearing for a day or so. On each visit he would assure me that we would be leaving soon.

My concern that "soon" wasn't soon enough, as the days morphed into weeks, didn't phase him. Likewise I had no success persuading him that I could not, and would not, be responsible for smuggling the nuns into Ethiopia . . . all to no avail.

And so I resigned myself to my fate that, added to the special cargo and my own precious skin, I would be responsible for two helpless middle-aged Irish nuns.

Accordingly, we whiled away the waiting days snoozing, swatting flies and loading up the 'landcrusher," as we called the specially-equipped Toyota Land Cruiser. Fortunately, the nuns had brought plenty of baggage for their year-long assignment, and considerable care was devoted to packing the cargo section in a way that should we be stopped and searched, there just might be enough goodies for sacrifice on the altar of bribery and corruption to spare the real cargo. Video cameras, and portable radios were popular bribes, but video cameras, though highly prized, could also lead to accusations of spying, something to be avoided at all costs.

The real top secret cargo – which no one other than my fearless Norwegian Project Leader Trygve in Khartoum and a few US Embassy personnel were supposed to know about– was a large red canvas bag with US MAIL in big bold letters stamped on it. The bag was about four feet high, with the girth of an NFL linebacker, or rugby prop forward, depending on your sporting preference.

It was cinched at the neck by a cord through brass grommets, and secured by a cheap combination padlock; nothing that a pair of scissors or a dull pocket knife could not have opened.
In it was seven million in cash.

SEVEN MILLION CASH IN A BIG RED BAG

Assembling several million in cash is not as easy as it sounds. You can't just walk up to the counter of the Ethiopian National Bank in Addis Ababa with a multi-million check from Uncle Sam and say, "Could you cash this for me in large bills, please?" without attracting awkward questions.

The 'grant' ** fell under Title 'some-number-or-other,' which specified soft, i.e. local currency, so as to preserve the 'plausible deniability' of not supporting the combatants. This of course means dollars must be exchanged on the black market. The local currency, the Birr, was about as much use outside Ethiopia as a bucket of Indian Wampas would be on Wall Street today. Never mind the fact that that's probably what Peter Stuyvesant paid for Manhattan in 1626!

But there was one place nearby where such magic could be accomplished . . . Mecca. Just across the Red Sea. The world's number one tourist destination, drawing millions of pilgrims annually, and the world's largest "grey market" currency exchange.

Perfect!

So with the aid of a Saudi friend or two, a network of street money changers, and a little patience the money started to assemble. It took a few weeks before enough Birr had been gleaned off the streets of Mecca and Medina, seven million in all, and now it was time to go to Jeddah to collect.

Now came the next problem: how to bring all that cash into The Sudan without attracting attention, and – Sudan's justice being notoriously brutal – how to bring it in legally, if at all possible. Seven million in used notes takes up a fair amount of space, even if bundled into nice elastic banded bricks of a hundred 100 Birr notes, as most of it was. All together, the loot filled eleven shoulder bags; pale blue shoulder bags with Citicorp on one side and Citicorp in Arabic on the other, courtesy of our Saudi friend; too much for carry-on and not bags that could be checked so commercial flights were out.

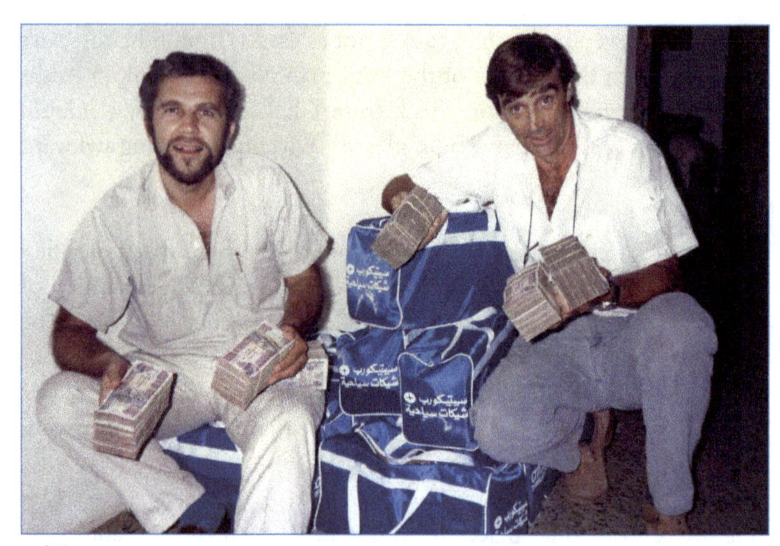

TRYGVE AND THE AUTHOR WITH ELEVEN PALE BLUE SHOULDER BAGS STUFFED WITH CASH.

That left a charter as our only option, and our Saudi connection came to the rescue with a private jet borrowed from a friendly Sheikh.

Where to land was the next issue, and with very few serviceable airfields – at any of which the arrival of a private jet would be unusual enough to attract attention – it was decided to land in Khartoum and take our chances. A few weeks prior, an attempted coup by Air Force officers had been brutally suppressed* and the whole country was on high alert and pretty jittery, which did not make things any easier.

However, some things don't change: the afternoons are very hot – a time for afternoon naps and a time for overload power outages, which shut down such air conditioning or fans as may make Khartoum Airport almost fit for human habitation.

So the plane was scheduled for a 3 pm. arrival, at the height of the doldrums. Technically, it was not illegal to bring money into the country, however all arrivals were required to fill out a currency declaration form with many lines and columns in which to record precisely the amount and denomination of all monies in the passenger's possession.

It was also well known that the customs officials were meticulous at counting every last cent, in the hope of discovering a mistake. Such errors led to accusations of smuggling and associated draconian penalties, which must then be negotiated away with a cash settlement.

With eleven shoulder bags all stuffed with cash, just counting the money would have been a nightmare involving every official, porter or janitor at the airport and then some. The only hope was the afternoon lethargy.

Accordingly, we had assembled about twenty different currencies. Knowing that the authorities were only interested in 'hard' currencies, the form was carefully filled out, starting with a few hundred dollars on line one, followed by a smaller number of dollar Travellers checks on line two, then Pounds sterling, Swiss francs, Deutschmarks, Belgian francs, and so on: one currency to each line, and each a smaller number than the previous line.

After the hard currencies, came the soft ones: Ugandan shillings, Malian francs, Congolese Zaires, Egyptian pounds, and so on: again, each number smaller than the previous until the very last line. Here the number 7 was boldly displayed, followed by the word million in a barely legible scrawl, then a clearly legible "Ethiopian Birr."

Fred, Trigve and I stood sweltering in the fetid air across from the customs officer. Beside us were, stacked in plain sight, the eleven shoulder bags, while the officer meticulously counted the various currencies presented and checked them off against the list. As he moved down the list, counting, verifying and checking off each offering, the tension mounted. We passed the 'hard' currencies and into the 'soft' ones, moving into the realm of trivial quantities and irrelevant values as the beads of sweat mounted on our brows.

We were now into the double digit numbers of essentially valueless currencies, and the officer was showing no sign of declining interest, when suddenly, with perhaps two or three lines to go, he waved his hand and stamped the document.

Gathering up the eleven bags we walked out of the airport to the waiting vehicle, trying hard to contain our exuberance.

Reporting to the Embassy the successful completion of the first part of the program, we were congratulated and issued with the big red US MAIL canvas bag. The next few days were spent hiding the booty in places like the trunk of Trygve's Mercedes, while we waited for the necessary travel papers. Travel within Sudan required permits, and following the attempted coup there were checkpoints everywhere.

Getting local travel permits didn't take long, typically a week or two. The Embassy was quite good at prodding the government, but the TPLF was a different matter. The Embassy could not deal directly with the rebels of a sovereign state, so we had to work through a surrogate: in our case, a Norwegian NGO based in Khartoum, called Redd Barna, that had operations in the Ethiopian provinces of Tigray and Eritrea.

The chap who provided the passes for travelling in these rebel-controlled areas was a mysterious one-armed Ethiopian named Abahdi, who lived on the outskirts of Khartoum, next to a large house surrounded by extra high walls topped with lots of razor wire.
It was the home of a Saudi Sheikh who had already achieved local notoriety for his lavish spending and radical Islamic rhetoric.

The Sheikh's name was Osama bin Laden. Need I say more?

With no contact other than a surprise visit, we dropped by once or twice a week mostly when he was out but occasionally catching him by surprise . . . Abahdi, that is, not Osama bin Laden!
When we did catch him, there was never actionable information, all things in the future being in the hands of the Almighty, "Insha'Allah" (God Willing) as the faithful qualify all things in the future, especially promises.

It did, however, help to keep our application on the front burner, so to speak, until finally the sacred permit arrived, and it was off to Kassalla with the loot..

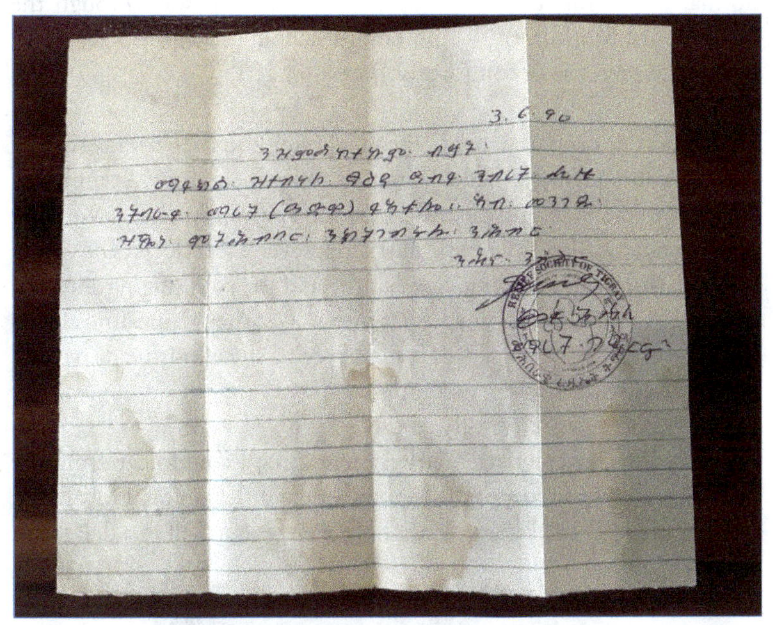

3.6.90

THE SACRED PERMIT FOR TRAVEL IN THE TPLF CONTROLLED AREAS.

In Kassala my contact introduced me to the Nuns and the waiting game started all over again, my new contact disappearing for days at a time and then reappearing to assure me that everything was going according to plan, with no relevant details, before disappearing again.

Finally, after what seemed an eternity, the day came.
Late in the afternoon, my Ethiopian – displaying an unprecedented urgency – announced that we must go immediately to the assembly point, and would leave at sunset. At the assembly point, there were 46 trucks, mostly beat-up old 8, or 10-ton Fiats and Bedfords, and at sunset we moved out, crossing the border a few miles later under cover of darkness.

We were placed in the middle of the convoy because of concerns about mines, which was comforting, even if it was more about the money than our lives. It was the dry season and the dust was incredible. Avoiding normal routes for fear of ambush, we weaved our way through the desert to the foothills of the mountains seldom exceeding 10 mph, to keep some pretense of comfort for the passengers.

The trucks cut deep trenches in the sand, often too deep for my smaller wheels; consequently I had to 'crab' in four-wheel drive, keeping my wheels on the side and not the bottom of the trenches. We kept crossing deep gullies full of fine dust often several feet deep. The trucks in front threw up blinding clouds of fine dust, so plunging into the gullies was more like driving a boat into waves, the talcum-fine clouds of dust exploding over our hood and covering the windshield so that I had to drive with the wipers running.

Every so often, after a particularly violent lurch I could hear, from the back seat, above the roar of the engine, little mutterings of "Holy Mary, Mother of God" or "Blessed Jesus, have mercy on us" all in a delightful Irish brogue.

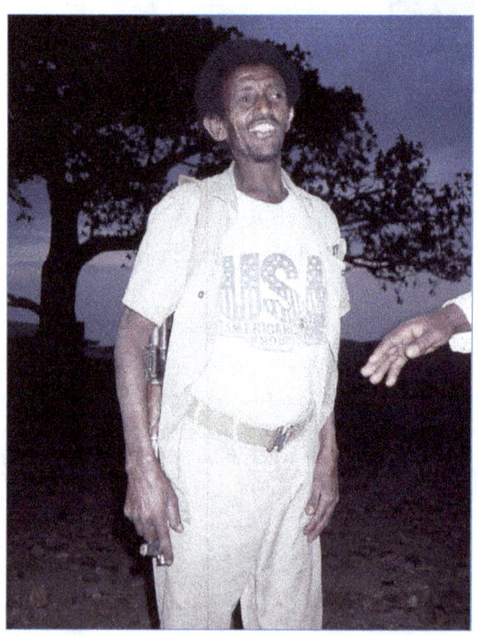

A WIRY LITTLE FELLOW CALLED SOLOMON, WITH ALL THE CHARACTERISTICS OF A DRILL SERGEANT

The convoy leader, a wiry little fellow called Solomon, with all the characteristics of a drill sergeant, had chosen my vehicle as his ride, presumably because it looked like the most comfortable of his 46 options. Thinking that he didn't know what we carried was not reassuring. It was not until well into the trip, when it became apparent that everyone knew the secret, that I started to relax.

Sitting beside me in this, his 'command vehicle' he would periodically pull us out of the line, and then – hanging out of the window and gesticulating wildly – he would yell at each truck, until the rearguard had passed, and he was satisfied that we were all present.

Then like a sheepdog, he had me overtake the entire convoy, bouncing over the terrain at breakneck speed until we were alongside the lead truck, where I had to hold station while he and the driver screamed at each other, above the roar of low-geared high-revving engines. All the while, the nuns and our heavily armed 'fighter' rattled around in the back seat like dice in a throwing cup.

We would then return for a brief respite in the relative comfort of the middle of the convoy before repeating the sheepdog routine.

All this, I should add, was done under blackout conditions. Our red tail lights were visible, but headlights were all blacked out except for a little slit that gave a glimpse of what lay a few feet ahead.

Around midnight, we pulled over in a wooded area for a break and a snack.

Dinner was Njira, which rhymes with injury, a thick brown goop with the consistency and appearance of axle grease, served on the flat lid of a cardboard box. Dipping a gob of Njira into a bowl of meat stew, with the right hand only, the nuns and I, closely watched by our native hosts, managed to overcome our reservations, and swallow our rations without grimacing.

Then it was back on the off-road again.

Sharing a bowl of Njira

In the pale light of predawn, we dispersed into a wooded copse and went to great lengths to hide mirrors and windows from glinting in the sunlight, lest they reveal our position.

Mengistu's MiGs prowled the rebel-controlled northern province of Tigray by day, looking for 'targets of opportunity' like us, which we were anxious to deny them. Fortunately, they had little or no nighttime capability, so virtually all movement and activity was conducted at night under cover of darkness.

The day was passed trying to catch some sleep: not an easy task in the heat of the day while beating off the flies.

SISTERS BRIDGET AND MARGARET
hiding up during the day

A couple of naps interspersed with cups of tea and another meal of Njira, and it was time to get ready to move out at sunset. This went on for several days and nights. At each stop, we left a couple of trucks, and the shrunken convey continued on to the next stop.

As we got deeper into the highlands the terrain got more rugged, and our progress slower. Added to which, as the days wore on, I began increasingly to experience the effects of sleep deprivation, a condition where hallucinations compete with vision, whether one's eyes are open or not.

TYPICAL MOUNTAIN ROAD,
NO PLACE TO HALLUCINATE

You have probably surmised, dear reader, that driving at night in the dust of a train of trucks on a narrow, rubble strewn, serpentine road cut into the side of a mountain with a rock wall on one side and a sheer drop devoid of barriers on the other is not the best place to hallucinate.

Somehow, with divine intervention, invoked, no doubt, by my two companion nuns, we made it...to the next stop that is, until finally – in some obscure mountain village whose name escapes me –the last trucks, Solomon and our AK-toting fighter abandoned us without explanation, just a wave and a "ma Salama" goodbye!

There we were, two Irish nuns and I, in the middle of nowhere, with no idea where we were, or where to go . . . not to mention a cargo of seven million in cash.

SOME OBSCURE MOUNTAIN VILLAGE
WHOSE NAME ESCAPES ME

Exhausted and devoid of inspiration, I closed my eyes and was out for the count instantly.

The next thing I knew, a young chap with a big grin, an AK, and several hand grenades hanging from a string around his waist was climbing into the seat next to mine. We established fairly quickly that he would guide us to our destination, which was fine . . . even though it was not clear whether he was our protector or a hijacker.

My problem was the hand grenades.

They were hanging from the string around his waist, threaded through the rings that pulled the firing pins.

Now for those who skipped grenade-tossing class, hand grenades look like baby pineapples with a spring-loaded handle or lever held in place by a pin attached to a pull ring.

Pulling the pin releases the handle, which in turn ignites a short, five-to-seven-second fuse: time for the tosser to throw it before it explodes.

Without Solomon, with whom I had a good rapport and language communication, I was reduced to negotiating in Arabic, our only common language.

Unhappily, his Arabic was even more limited than mine, and to make matters worse, my accent was Yemeni and his was Sudanese. Imagine, if you will, an argument between two immigrants with limited English, one who learned his English on the streets of Brooklyn, and the other in the bayous of Louisiana... you will get the picture.

Worse still, beyond the ubiquitous AK 47, wearing a bunch of handgrenades around the waist was a mark of status second only to carrying an RPG, which fortunately would not fit inside my vehicle. Against this challenge to his virility was my paranoia at the thought of an activated hand grenade rolling around the floor in the dark after a particularly violent bounce in the road, its pin still hanging on the fighter's belt.

Eventually we reached a compromise, agreeing to a waiver of the 'no hand grenades' rule, allowing our fighter-guard-guide to keep his hand grenades but under the seat, his seat . . . with no strings attached to the pull rings!

The following morning, we arrived in Adua, our final destination, and I was able to discharge – to my great relief – my responsibilities for the nuns and the cash. It was only then that I was able to disclose to them the nature of our cargo. The gasps and the giggles and the invocations to the Almighty and His immediate family defy description.

When the giggling finally subsided, I was able to persuade the nuns to pose with AKs in front of the stash.

May God forgive me! Though I have located photos taken at the time, that particular photo is missing... perhaps the work of a celestial kleptocrat.

Footnotes:

*The leader of the coup was a well educated Air Force officer from a prominent family in a powerful tribe. When the coup failed, the officer negotiated a surrender to avoid further bloodshed, knowing that he would likely be executed after a trial. Instead, he was summarily shot upon his surrender. His execution without trial – and worse, on a Muslim Holy Day – was far more offensive to his family and tribe than the execution itself.

Later we heard that not long thereafter an entire central government committee was assassinated while in session.

The News was suppressed and the perpetrators never prosecuted. Apparently in the world of an eye for an eye and a tooth for a tooth Honour was satisfied and justice done!

**Background to the Farm Price Subsidy Program, the official title of the project to which I was assigned. In the mid 1980s, drought, exacerbated by civil war, led to the mass migration of the rural population of Tigray, the mountainous Northern province of Ethiopia.

The Exodus was well organised, as village after village faced with starvation in situ or survival elsewhere migrated North en masse into Muslim Sudan. I say Muslim Sudan not to foster religious conflict but to explain just what a dramatic decision this must have been for people who were still in a state of religious fundamentalism not unlike medieval Europe with little if any concept of the secular rationale of we Westerners. They saw themselves as both the Original Church and an Island of Christendom beating back endless waves in an Ocean of Islam. Among them were many thousand of the Falashas, the black Jews of Ethiopia, which calls for another digression, so bear with me.

In Ethiopian tradition the Queen of Sheba, who ruled a great and prosperous nation that supplied the temples of the ancient world with precious Frankincense from the indigenous Gum Arabic tree, paid a visit to the other super power of the region, King Solomon of Israel. Apparently the visit went beyond diplomacy and the pregnant Queen returned home with an entourage of Jewish retainers to raise the Royal Offspring appropriately for his future responsibilities. The child was Menelek the First, the Lion of Judah and the King of Kings, a title worn by every Emperor of the region, whether it was called Abyssinia or Ethiopia until the deposition and Murder of Hailie Selassie in 1974. The Falashas or black Jews of Ethiopia claim descent from that original entourage afforded the Queen of Sheba by King Solomon and now numbering in the many thousands, joined the mass exodus of Christian Ethiopians into Shariah Sudan.

This caused panic in the intelligence community (really!) and triggered a black ops extraction of the Falashas back to the promised land from whence they had departed three thousand years before.
But that's a story not yet ready to be told.

Back to the point, the Ethiopian famine attracted global attention with chaps like the Irish punk rock star Bob Geldof and his Boomtown Rats who created Band Aid, raised Millions and got the World's attention big time, and with whom I had a couple of interactions, but that's yet another story.

To forestall a repeat performance, the International Aid Community decided to stop the migrations by feeding the starving in place, and numerous agencies started delivering food directly to the stricken areas whether stricken or not.

As usual the Law of Unintended Consequences kicked in.

FOOD DISTRIBUTION IN TIGRAY

Unable to compete with free food, the farmer quickly discovered that he could send the wife to the distribution centre, usually a day's journey away, to collect their rations and he could stay in bed.

If he was a kind and generous husband he might even let her take the donkey so she didn't have to carry the sack of grain on her head.

Why not? If he didn't need the donkey to plough the field to plant the seed that might yield a crop to feed his family and the donkey, why bother. So farm prices collapsed and the centuries old terraces that retained soil and water for hillside agriculture were not maintained and disintegrating.

Our mission was to purchase grain from farmers at a price to keep them in business, and add the grain to the 'free' distribution program managed by the TPLF. In addition to sustaining the local farming industry, it was likely a way to support the revolution with hard currency while maintaining 'plausible deniability'; laundering the Dollars into Birr, the Birr into grain, and then buying back the Birr from the TPLF for Dollars 'on the street' in Mecca... but nobody wanted to go there!

A Tale of Two Nuns: Part II

Sudan/Ethiopia 1990

Some weeks after arriving in Adua and discharging my responsibilities for the cash and the Nuns to other parties, I found myself late one night in the village near Margaret and Brigitte's Convent. We only moved at night to avoid becoming targets of opportunity for Mengistu's MiGs. so I'd got used to being out and about at night and had forgotten that most sensible people still went to sleep at night. In addition we had been on the road for some time, a convoy of two or three vehicles, our TPLF escort the diminutive Teklawaynie, his deputy and interpreter Heshai and a dozen or so heavily armed 'fighters'.

The diminutive Teklawaynie with shirt always untucked and Heshai left.

The lads wanted a break to catch a meal, visit girlfriends or whatever in the village and I wanted to say hi to my Nuns, so it was agreed that we would rendezvous in an hour at the Convent gate.

Accordingly the lads drove off into the town and I banged on the big metal gate without pause to consider how inappropriate let alone how inconsiderate I was being. To make matters worse I had not shaved for a couple of weeks nor had the chance to don clean clothes for a few days. After a minute or two making quite a racket banging on the metal gate a voice from the other side asked who was there. I explained who I was and said I had hoped to pay my respects to Sisters Brigitte and Margaret whereupon the gates opened to reveal a Nun in her full regalia clutching a crucifix who I shall call the Mother Superior. She stared at me wide eyed for a moment, I could not tell whether in horror or wonder, and then said: "You're the man who brought us Brigitte and Margaret". There was an air of reverence in her voice that I found intimidating, not wanting to be mistaken for a divine apparition, I started to apologize for the intrusion, the late hour and my unkempt appearance, but she would have none of it. "Ye must be 'hongry,'" she said, in that delightful Irish brogue, leading me across the courtyard.

"Oyl fetch dem, dey'll be trilled" and the next thing I knew I was led into a room with a very long table running its full length, and set down at its head. The room was more of a hall than a room and could have been the setting for a painting or a film set of the Last Supper. Devoid of windows the single long wooden table ran the length of the room with wooden benches on each side and a sort of throne-chair at the head in which I had been seated.

Along the walls were life-size portraits of saintly figures in pious poses, palms together, peering benignly down at me from beneath their halos. A couple more nuns arrived and set some buns and a pot of tea in front of me before stepping back, hands clasped together to watch me eat. Then the church bell started ringing and soon more nuns arrived, in twos and threes, some with food offerings and some just to stare at me it seemed, and watch me eat.

Before long there was a semicircle of a dozen or more nuns standing around me as I struggled between mouthfuls to explain how I came to be smuggling large quantities of cash and Irish nuns into rebel controlled Tigray in Northern Ethiopia; there was really no simple explanation!

Then Brigitte and Margaret arrived and a wave of giggling started as they came over and hugged me... apparently such public displays of affection with a strange and scruffy man were too exciting for words and everyone started giggling like naughty schoolgirls, even the mother superior was convulsing a little as the infectious giggling spread.

The bell ringing in the middle of the night had alerted the neighboring monastery and soon the numbers swelled as monks from across the wall started to show up in their brown Habits to watch the lone diner. It was clear that the stories of their journey had become part of convent legend and with such a rapt audience I couldn't resist playing to the crowd with a little of what one might call 'Homeric Embellishment'. I told the Mother Superior and the entire assembly that Brigitte and Margaret were two of the finest smugglers that I had ever had the privilege to work with... which was met with gasps and giggles in equal measure at about the same moment as the honking of an old Toyota Land Cruiser and, as I recall, a short burst from an AK47 signaled that my guys had come to take me away!

And so ended the Last Supper.

Spontaneous Demonstrations

Tigray, Northern Ethiopia, 1990 / 1991

"Wake up ! Wake Up! We have to go! NOW!"

The urgency of the voice awoke me instantly from a deep 2am slumber, and I was out of bed and into my jeans, both legs at once and racing out into the night after the voice, my guide/guard and interpreter, the diminutive Teklawainy. Certain that a midnight raid by the enemy – the government forces of Colonel Hailie Miriam Mengistu – was imminent, I wasted no time diving into Teklawainy's beat-up old Toyota Land Cruiser, as he barreled off into the dark night, heading for the mountains.

Propaganda portrait of the enemy Col. Mengistu

After an appropriate interval to catch my breath and allow my heartbeat to settle back to near normal (not to mention being a safe distance from the town, Adua), I ventured to ask "Where are we going?"

"We are going to Makele, to see spontaneous demonstrations tomorrow morning at 9 o'clock," said the diminutive Teklawainy in a matter-of-fact voice.

Makele was one of the regional towns that had recently been taken over, or 'liberated' to use the preferred terminology of the TPLF* rebels, and lay a good 5 or 6 hour drive from Adua through the mountains.

The road was tortuous, unpaved and full of hairpin turns, as it threaded through brutal terrain with sheer drops on one side and rock walls on the other, all of which made it a pretty hazardous journey at night.

The road through the mountains to Makele

With nowhere to hide in daylight, vehicles were sitting ducks for Mengistu's Migs, which roamed the skies by day looking for targets of opportunity, thus making nocturnal travel the only option. And so it was that around 8am we successfully emerged from the mountains and headed down into the plain and the town of Makele.

Ongoing negotiations between opposing entities are a fact of life in wars, especially civil wars, and since most governments refuse to recognize their rebels, there remains plenty of room for intermediaries: typically UN agencies or NGOs of standing, such as the International Red Cross (ICRC).

In Ethiopia at the time, the lead Agency was the Ethiopian National Church, since most of the people on both sides professed to be Christian. Ironically, the leadership of both sides were professed atheist Marxists. The Soviet-backed Mengistu Regime espoused the Soviet version, while the TPLF claimed to be true Marxists of the Stalinist or Albanian Enver Hoxha variety, battling the Soviet Revisionists!

Arriving in Makele shortly before the 9 am kickoff, so to speak, we drove slowly down the main street, pausing briefly at every side street, to see the lines of 'Spontaneous Demonstrators' all assembled with placards, each line under the control of a man with a bullhorn waiting for the signal to march.

At each stop, Teklawainy identified the group. There was the Teachers Union, the Farmers Cooperative, the Laborers Union, the League of Cotton weavers, this Association and that Organization, and on and on, all demonstrating for the TPLF and against the 'Derg', the Soviet backed, Mengistu Government.
All were in position and poised for the signal to march and echo the bullhorn chants.

At the end of the main street was the town square, all set up for the occasion with a raised dais, or platform, for representatives of the Church, the Government and the TPLF. . . and of course all the world's News Agencies, ABC, BBC, CBS, all the way down to PBS. (the networks were prioritized alphabetically, lest bias be inferred), and strategically set up with TV cameras to broadcast the event in real time.

It was the media presence, a rare event in Ethiopia in the days when the cyclops eye of the global media was briefly distracted from its single focus: the big stories of Afghanistan, Kuwait and Iraq, that had triggered the frenzy. So we spent the day in Makele watching spontaneous demonstrations and stayed the following night to digest the experience.

Not wanting to risk a daylight return through the mountains, we spent the next day exploring the town and learning of the problem of mines associated with taking over strategic towns. Guerrilla warfare relies on stealth, and minefields are an effective way of restricting stealthy intrusions into urban environments. In Asymmetric warfare, minefields are increasingly focused on maiming rather than killing; wounded warriors (in compassionate cultures) require vastly greater human resources than dead ones.

Sorry to say but it's true.

So we learned that there were large areas around the town seeded with mines. Not big mines designed to blow up vehicles or tanks (which you can hear coming anyway). The Soviet specialty at the time both in Afghanistan and Ethiopia was little mines designed not to kill, but to maim. Designed to blow off a foot or a hand, and sadly often designed like toys to attract curious children, and blow off a hand.

Lacking the resources for expensive minesweepers, the locals used goats. Scattering corn in front of a herd of goats is a cheap and effective minesweeper. Not pretty for animal lovers, but at least the casualties provided dinner.

That evening, after 'dinner' we headed back through the mountains to Adua and ever since I have been fascinated by the miraculous phenomenon of 'spontaneous demonstrations'...

*__TPLF__: Tigrayan People's Liberation Front.

The Head Rebel (Meles Zenawi)

Adua, Tigray, N. Ethiopia 1990

Meles Zenawi, President of Ethiopia, died August 20, 2012.

I first met Meles Zenawi in Adua, Northern Ethiopia, in 1990. At the time, he was the head rebel: Leader of the TPLF, the Tigrayan People's Liberation Front, waging a brutal civil war against the Soviet-backed Marxist Regime of Col. Hailie Miriam Mengistu.

The Cold War was still raging, though the fires that fueled it were waning, with Gorbachov, Perestroika and the withdrawal of Soviet forces from Afghanistan, as the Soviet Union prepared to disintegrate.

The TPLF represented themselves as 'true' Marxists, unlike the 'revisionist' Soviet puppet Colonel Mengistu, and had effectively taken control of Tigray, the Northern province. Although the TPLF controlled the ground, the Mengistu Regime still had control of the air, and by day, Ethiopian Migs strafed and bombed targets of opportunity at will, making travel, or assembly in the open, hazardous.

Consequently almost all activities were conducted at night. Needless to say, Meles was a high-value target, and had already survived several assassination attempts, near misses you might say, which served to enhance his charisma.

He would just appear in the middle of the night, have a chat with the TPLF chaps and then disappear into the night again.

This particular night, there were three or four local TPLF lads; Fred Cuny, a well-known American Relief Specialist; a couple of chaps from the Embassy in Khartoum; and myself, in a room best described as a cross between a mud hut and a bunker. Uncle Sam had been helping the rebels for some time in ways careful to avoid accusations of participation, wherever and whenever the opportunity to discomfort the Russian Bear or his surrogates arose.

But now, with Gorbachev and Perestroika, the possibility that the TPLF and their Eritrean friends next door – the ELF* – might actually prevail, had cast a whole new perspective on things. With the prospect of a much higher profile on the world stage, not to mention U.S. recognition of a new regime, the emphasis changed from Russian bashing, which was all good clean Cold War fun, to 'OMG, we might have to invite these guys to dinner at the Whitehouse.

So polishing up their act became the order of the day, and, I suspect, was Fred's directive. Dropping the Marxist Party line and especially referencing the Stalinist or Albanian models were now high priority.

Being low on the totem pole. I was not invited to address the great leader until after the sensitive tactical and political phase of the discussion was over. By which time considerable quantities of the local beer, called 'sewer,' had been consumed, lowering the gravitas of the occasion. The brew didn't come in cans or bottles but was scooped out of a large earthenware jar in which the original fermentation had been accomplished, and the decomposed ingredients remained.

Consequently, when it came to my turn to ask a question, I was at a loss, all the relevant questions had been asked. So I inquired, "How did you get into this unusual line of work, making revolutions?"

Meles burst out laughing, while the assembled audience gulped hard.

For the next twenty minutes he told the story. It went something like this: "As a first year engineering student at Asmara University, and being among the first group of students from illiterate peasant parents to reach university since Haillie Selassie opened the door to universal education, he had joined the Students Union and become an active participant.

The Union was rife with political passion, and a debate on the policies of the Mengistu regime, which had overthrown Haillie Selassie, led to a resounding vote of rejection, following which the question was asked, "So what do we do now?"

One of the students, brandishing Mao's Little Red Book, proposed a motion 'to take up the armed struggle'.

The motion carried, so a bunch of them dropped out and took to the hills to start the revolution.

Unfortunately, the peasants were not interested in politics, nor especially in being lectured by young students.

Frequently a peasant, recognizing one of them, would ask "aren't you so and so's son?" Or "why aren't you at school? Go home and milk the cows, or help your family plough the land. Do something useful and not this politics nonsense."

Things were very bad: they had no money, no food, no guns, and the people were not interested.

You can't make revolution when you have no money, no guns and the people are not interested in your ideas.

Then one of the lads had an idea.

He had once driven the ambulance in a nearby town, and knew where it was parked, and – more importantly – where they kept the key. So a small group of the lads snuck into town late one night, stole the ambulance, drove it into the local police station, sirens wailing, jumped the sleepy guards, stole their guns, and then knocked off the bank.
Back in the hills, they now had money and guns, so things went a lot

better, but still the peasants were not interested in politics or their revolution.

However, they were able to cause a lot of mischief by attacking government offices, police stations and the like, hit-and-run style.

Some of their raids were pretty audacious and the chief mischief-makers became known as the Firecrackers, earning a degree of celebrity, at least among some who got a chuckle out of their audacity.

Little by little, they built up a small network of followers, sympathizers, spies and such, but the peasants and farmers, fearing retribution from the Government, still resisted joining openly.

It was not until the Mengistu regime, frustrated and embarrassed by these relentless pinpricks, resorted to brutal reprisals on civilians, known as the Red Terror, that the people started to respond to the revolution and join the TPLF.

The Red Terror counterinsurgency program was taken from the Nazi Gestapo and refined by the Communist East German Stasi under the Soviets. In Soviet-controlled surrogate states, it was the East German Stasi who trained the local secret Police.

Ethiopia was no exception.

Knowing that the core of the insurgency was in the University, they would round up random students or faculty, and accuse them. Under torture, they would extract two or more names, who in turn were arrested and forced to name others.

It didn't matter whether the arrested were involved or not. The theory was that sooner or later, the interrogations would break into a cell. Those who resisted or confessed were often executed, and their bodies, wearing an accusatory placard, were dumped on the sidewalk outside their homes to rot. It was illegal to bury or in any way cover the corpses.

This barbaric practice, designed to cow the populace, had the opposite effect. It proved to be a powerful recruitment for the Rebels, who with a little help from their 'friends' gained control of the northern highlands, and advanced steadily south. When the Cold War ended, without Soviet support the Mengistu regime collapsed, and Meles Zenawi, the Head Rebel, became President, and later Prime Minister, until his death in 2012.

*__Eritrean Liberation Front__ (ELF)
A war of succession that had raged in varying degrees of intensity since the 1960s, and was finally consummated following the overthrow of the Mengistu Regime by their Allies, the TPLF, in 1991. Eritrea is the coastal strip of Northern Ethiopia on the Red Sea, which the brief Italian colonization in the 1930s developed significantly over the highland hinterland of Tigray.

Their mutual cooperation against the Soviet Mengistu Regime fell apart a decade or two later with a military incursion into Tigray by the now independent State of Eritrea.

Funny how allied Autocratic Regimes become adversaries without a common foe.

A Field of Bones

Shiraro, Northern Ethiopia 1990

Shiraro is a nondescript little town in the highlands of the northern Ethiopian province of Tigray. It nestles in a valley between rugged mountain ranges, with essentially only one way in or out through a tortuous narrow access road.

Overlooking the town is a flat-topped mountain, sort of like a butte: very steep sides and a flat top.

During the civil war, Shiraro was the garrison town of the Ethiopian Third Army, a thirty thousand man Soviet-equipped force of the Soviet-backed Marxist regime of Colonel Hailie Miriam Mengistu.

The TPLF rebels were able to shut off Shirraro's lifeline by ambushing the re-supply convoys in several 'choke' points where they could take out the lead and rear vehicles with an RPG, and then finish off the trapped convoy at their leisure.

For a while, the garrison was maintained by an air lift, until the rebels were able to capture the flat-topped mountain overlooking the town and its airstrip.

I was taken up to see the battlefield for control of this strategic high point.

The road to the top was a tortuous series of hairpins, a veritable suicide march for a column of fighters, so instead they scaled the almost sheer sides of the mountain to gain the flat top, where the defenders had set up deadly fields of fire between the bunkers and redoubts. The whole area was littered with bleached human bones and shell casings in almost equal numbers; a stunning testimony to the ferocity of the battle.

Having lost the high ground, the airfield below became unusable; and the town – cut off from resupply – was untenable, so the garrison army had no choice but to evacuate.

Next, I was taken along the evacuation route, a trail of devastation as the column battled its way through serial ambushes at every bottle neck of the escape route.

REMAINS OF THE ETHIOPIAN THIRD ARMY
RETREAT FROM SHIRARO

The escape route was the road to the Takezzi Narrows gorge, about 20 or so kilometers from the town. The gorge is a smaller version of the Grand Canyon in Arizona, carved deep over the ages by the Blue Nile or one of its tributaries.

Crossing the gorge involved descending a tortuous series of hairpins down the cliff face, to ford the river and then up the other side again, via another series of hairpins. It was here that the TPLF rebels had set up the killing ground and where the surviving remnants of the Ethiopian Third army – now mostly on foot having lost their vehicles and equipment on the way – surrendered.

Twelve thousand prisoners were taken, all that was left of the original complement of thirty thousand.

The TPLF had neither resources nor interest in holding prisoners. So captured government troops were given the option of going home, joining the rebellion or just 'hanging out' in Tigray. This worked extremely well and actually encouraged the defection or surrender of Government troops, for whom going home would inevitably result in being re-enlisted in the army and possibly facing desertion charges.

The only ones who were held against their will were a few with technical expertise for maintenance and training purposes of weapons whose deployment was beyond the capability of the peasant boys and girls who made up the bulk of the rebel forces.

For political reasons, the official story was that the rebels relied on captured weapons to sustain their rebellion. This was obviously nonsense, but at the time it was necessary to supply the rebels with Russian-made equipment so as to preserve the myth.

It was not until President Ronald Reagan introduced the American-made Stinger missile to the Mujahideen in Afghanistan that the myth was exploded.

Besides, there was plenty of stuff from all sources available in the great International Arms Bazaar . . . for a price, of course!

Later it became apparent that our "farm price subsidy" program was quite possibly a source of hard currency for their war effort; but that's another story. . . (See A One-Armed Ethiopian named Abahdi . . .)

A One-Armed Ethiopian called Abahdi

Ethiopia, 1990

Getting a permit to travel from Sudan into Ethiopia was a problem. The process itself was not that complicated but the time it took was the great unknown. We, the functionaries, would apply and then wait for days, weeks or even months for the issue of the sacred

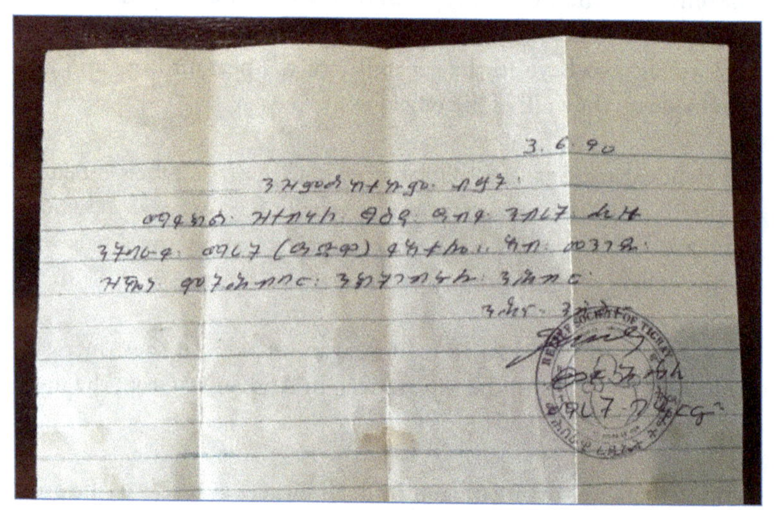

3. 6. 90

The Sacred Pass for travel in Tigray

The pass for travel within The Sudan took days or weeks; the travel pass from the TPLF, the Tigrayan People's Liberation Front, for the war zones of Northern Ethiopia and Eritrea took weeks or months. The issuer of the TPLF pass was an elusive one-armed Ethiopian named Abahdi, whose office was on the outskirts of Khartoum, and since there was no other way to contact him, we would 'mosey' over there every few days to enquire, usually unsuccessfully, as to the status of our application.

I remember that the office was near a residence surrounded by a high wall topped with lots of razor wire, the home of a rich Saudi sheik by the name of Bin Laden whose lavish spending had made him somewhat of a celebrity both on the local and national levels.

The protocols during the Cold War required that humanitarian relief be scrupulously neutral; this meant that any and all assistance on either side of the conflict be non military. Through the Roman logic of 'reductio ad absurdum' this meant that not only could we not provide the weapons of war but also the means to procure them. In consequence USAid funds* had to be provided in 'soft' or local currency that had no value on the world market. Needless to say much ingenuity was applied to bending the rules without actually breaking them.

To forestall another massive famine induced migration like the one in the mid eighties, the UN and various other agencies and NGOs had initiated free food programs, a sort of international home delivery food service.

FREE FOOD DISTRIBUTION, TIGRAY, N. ETHIOPIA.

Note the guard's footwear crafted from old tires.

Needless to say, the free food had collapsed prices in the market. Farmers were going out of business sending their wives to the nearest food distribution center rather than rising at dawn, harnessing the oxen, plowing, sowing and hoping for a crop and a return on their efforts at the market. The kind and thoughtful farmer probably added: "Oh! And Honey, take the donkey so you won't have to carry the sack on your head."

PLOWING AND SOWING IN THE ETHIOPIAN HIGHLANDS

Worse still, the labor-intensive terrace walls, developed over centuries to contain precious topsoil and water, were not being maintained, with catastrophic and irreversible results.

Like a dam, once breached, the floodwaters sweep it all away carrying precious soil down the mountainside to the river and out into the oceans.

TERRACE FARMING, THE AUTHOR WITH A LOCAL FARMER

Our mission was to subsidize farmers by buying their crops with a US $6 million grant from USAID, which first had to be converted into local currency to preserve the myth of neutrality.

Exchanging $6 million US dollars for Ethiopian birr could not be done at the Ethiopian National Bank without raising eyebrows, so it had to be done "on the street" . . . but that's another story.

Suffice it to say, nothing was quite as simple or straightforward as at first it seemed. About the only place in the world with a 'gray-market' currency exchange on this scale is Mecca, to which millions of pilgrims from all over the Muslim world travel every year.

So that's where, with the aid of some friendly Saudi bankers, the exchange was accomplished.

Apart from smuggling the cash into Sudan – a risky enterprise that was successfully accomplished, we achieved a bonus by scoring an exchange rate of 6.8 to the dollar, while other NGOs who entrusted the money-changing to the TPLF were getting only 3.5, much better than the official exchange rate of 2.8, but nowhere near the bang for the buck that we scored!

Now it did not take long to realize that someone had been skimming a healthy 40% off the exchange transactions until we came along and 'queered' the pitch, as they say.

We also ascertained that the 'market' had been floating around in the 6.5 - 7.0 range for a while, so it did not take a genius to realize that we were treading heavily on someone's toes!

Whose toes, and how would they react was the question. . . and I don't mind admitting that I was sure that some misfortune, quite possibly fatal, was a distinct possibility to get us out of the game...not a comforting thought. Nevertheless, the Rubicon had been crossed, the die was cast and there was no turning back.

Happily, all went well and the goods were delivered without mishap. (See: A Tale of two Nuns).

My role now changed from smuggler to monitor, and daily we visited farmers who showed us bags of grain for which my TPLF escorts paid them cash, from my recently delivered stash of Ethiopian birr.

Whether the grain was home-grown or supplied by USAID, Norwegian Church Aid or whomever, and re-bagged into local sacks, there was no way to know . . . and one did not pursue such thoughts, at least not openly. For that matter, I could have been shown the same stuff being moved from farm to farm ahead of our visit . . . "Ours not to reason why, ours but to do or die" as Rudyard Kipling or someone once astutely observed.

One thing quickly became apparent though. The price of grain started to rise rapidly, as did the cost of domestic transport, for which Uncle Sam was also paying.

By the time we had spent half the funds, the price of everything had risen by about 40%. The usual culprits were invoked: inflation, the war, shortages of everything etc.

(Interestingly, as I write this 30 years later, the same excuses are being officially invoked in the US and blamed on the war in Ukraine!?)

But back to the point, there was clearly something else going on. My suspicion that there might be a connection between the loss of 40% at the currency exchange end and the 40% price increase at the distribution end, was actually quite comforting. If it was the TPLF – and it was hard to think who else it might be – then I could stop worrying about 'interference.' After all, they were still getting their 40% slice of the pie, and we were incurring the risks and hard work of money-changing and smuggling, all of which incurred costs, whoever did it.

On my return to Khartoum to debrief and write my report prior to the next run, I contacted our Saudi banker friend who coordinated the money gathering in Jeddah, and asked him casually if there were any other players in the market for Ethiopian birr. "Not really", he said, and then after a reflective pause, "There is one chap who comes in every month or so, a one-armed Ethiopian named Abahdi, but he's usually selling birr."

I submitted my report with this tidbit of information and was promptly told to redact that part. . .

We did a total of three runs without interference. Looking back with the benefit of hindsight, I can't help but wonder whether the whole thing was a set-up by those in high places to insulate themselves from accusations of breaking the law.

In which case, it would have been nice to have been tipped off, at least with a 'wink wink, nod nod, know what I mean?' hint. Instead, we were

left to sweat it out.

Maybe Sir Humphrey Appleby was right when he asserted that the Official Secrets Act is to protect Officials, not Secrets.

From Tigray to Typhoons

Ethiopia 1990-1991

On the second money-smuggling trip into Tigray, my companion was a delightful Jesuit priest, Father O'Sullivan. We had met briefly on my previous trip, and bumped into each other again at the Khartoum Airport, where he was returning from the Vatican and I from somewhere else.

Quickly establishing that I was heading back to Tigray, he asked if he could ride with me. Of course! And the deal was done.

A splendid Irishman who had already spent a quarter century in the region, he was fluent in all the local languages, a local historian, and all-round academic, who literally knew everyone who was anyone, and then some, there being very few institutions of higher learning outside his Catholic seminary.

His alumni included Loyalists and Revolutionaries of every stripe and persuasion, and, miraculously, he had managed to stay above the fray, commanding respect from one and all.

Apart from sheltering under his protective wing, Father O'Sullivan was known and revered everywhere we went, he was great company to boot!

He was returning from Rome, where he had attended the now-famous Three Tenors Concert in the Colosseum, where Pavarotti, Placido

Domingo and Jose Carreras celebrated the return of Carreras after a bout with Cancer. Better still, the Father had a cassette recording of the occasion, which sustained us through the ten-day nocturnal journey.

Despite the recording's inferior quality, volume control prevailed over the competing diesel engine noise. Nessun Dorma will never be the same again.

Not long after this journey, when I had delivered Father O'Sullivan to his seminary, I found a little hole-in-the-wall restaurant that served fresh greens and salads . . . sort of. The proprietor was the son of an Italian immigrant from the Mussolini era who had succumbed, understandably, to the charms of a local lady, and elected to stay, when the Italians went home.

After weeks subsisting on bread and meat in a culture that despised green vegetables, the prospect of a fresh salad was irresistible. Soon after, without getting too graphic, I started passing coffee-colored water and, well, yellow diarrhea. More serious was a total lack of energy. Exhausted, I would take a nap and wake up fine, only to be knackered again ten minutes later.

I hung in for a while hoping the condition would go away, but it didn't.

Worse still, I was starting to get bouts of paranoia, something I had never had to deal with before. With no medical facilities, period, I had to get back to Khartoum. I was in no shape to drive at night, and off-road, both of which require an exhausting level of mental and physical exertion. It would have to be Heshai, my interpreter. Only small problem: he didn't know how to drive!

Heshai desperately wanted to learn how to drive, but despite several attempts at putting him behind the wheel, he was no match for the terrain. The prospect of hitting a rock, a rollover, or any one of a number of accident scenarios were not worth the risk.

HESHAI, MY INTERPRETER
wearing a Mucky Duck hat from Captiva Island, FL.

Somehow, we made it back to Khartoum, where I drove straight to the Hilton and consumed a large steak bearnaise before going to the Embassy to see the doc, who took one look at me and laughed, adding that the steak bearnaise was the worst thing I could have done.

"Want the good news or the bad?" he asked.

I chose the bad. "You've got hepatitis," he replied. "Look at your eyes, they're yellow."

I hadn't had the luxury of looking at a mirror for some weeks, so how could I know?

"The good news is it's almost over," he added.

It was that salad! I had survived for years by never drinking unboiled water – even ice cubes, not that we saw those very often – but that fresh salad was washed in unsterilized water!

After returning home to recuperate, I guess Fred felt sorry for me and sent me on a cushy mission for a change: a typhoon mitigation study for FEMA, (the Federal Emergency Management Agency).

There's a string of exotic atoll Islands in the Pacific, which the U.S. liberated in World War II from the Japanese, who had liberated the islands from the Germans after World War I. The islands run from the Marshalls, along the Eastern Carolinas (now called the Federated States of Micronesia), just north of the equator, to the Marianas in a giant flat U.

Tropical storms, whether Atlantic hurricanes; Pacific typhoons; or Cyclones, in the southern hemisphere; are the same thing. They originate in the Inter Tropical Convergence Zones (ITCZ) which lie in latitudes 5 to 15 degrees on either side of the equator, and trigger storms when the ocean temperature exceeds 80 F.

The Islands were being liberated again, this time from Uncle Sam's benign custody to independence as the Federated States of Micronesia. Lest they be acquired again by an unfriendly power, Uncle Sam left a number of welfare programs in place, presumably to make them too expensive to acquire. One of these was FEMA.

Originally set up to provide continuity of government in the event of nuclear war, FEMA (which allegedly has vast underground bunkers) morphed into disaster management of the domestic variety: floods, earthquakes, hurricanes, tornadoes and whatever else qualified as disaster . . . but always confined to the homeland or trust territories.

Our mission was a mitigation study so that FEMA would know how best to respond to a disaster.

Apart from the occasional and rare tsunami, it was all about typhoons, the Pacific Ocean annually spawning a hundred or more, compared to the Atlantic's dozen or so.

And so, in the fall of 1990, it was off to the Pacific to chase typhoons.

CHAPTER 6:
Chasing Typhoons

Chasing Typhoons

1990-1991, Micronesia.

(Formerly U.S. Trust Territories of the Eastern Carolinas.)

Micronesia is a collection of Island Archipelagos. Atoll Islands buffered from the ocean swells by coral reefs around a central lagoon.

Long dormant or extinct volcanoes, rising from the sea floor, formed the seamounts upon which these exotic islands, like strings of pearls, emerged in the translucent turquoise waters of the tropical Pacific.

Typically the majority of the population lives on a single Island, the capital and the seat of government, a coagulation which brings with it all the commerce, conveniences and crowds that clutter the modern world at the expense of the beauty, simplicity and tranquility of the unsophisticated out Islands, where dwell the other ten percent... and the occasional refugee from the crowded conurbations of the modern world . . . usually in an island hopping live aboard sailboat.

The Marshall Islands

Bikini, Enewetak, Wotho and Kwajalein 1990-91

From Honolulu, Hawaii, our first stop was Majuro, capital of the Marshall Islands, and by far the most populous of the Marshalls. Its population was boosted by the inhabitants of Bikini and Enewetak Atolls, which, after being liberated from the Japanese after WW2, were evacuated and used to test atomic bombs.

Bikini atoll enjoyed 23 nuclear detonations, making it the 'hottest' place on earth, and inspiring a Frenchman to launch his 'hottest' ladies swimwear, the Bikini. Unlike the atoll, the fashion bikini has yet to cool down.

Enewetak had slightly fewer bombs, but mostly bigger ones, as thermo-nuclear (or "H") bombs became the rage in the early '50s. As far as I know, it has yet to inspire a line of ladies beachwear called the 'Enewetak.'

The original Bikini evacuees, some 40 families totaling under 200 souls, adapted effortlessly to beer and welfare on Majuro, and have multiplied to an estimated 5,000 now. Unfortunately, nuclear test sites did not qualify as disaster zones, so we were not able to visit either. However, one of our group, Leo Migvar, an agronomist by profession, had made frequent visits to Bikini, and was able to satisfy, vicariously at least, my curiosity.

He described the Islands as flourishing, as though nothing had happened, with rich and abundant flora and fauna, a testimony to nature's resilience. The handful of officials stationed there wore radiation patches and were rotated frequently. But the former inhabitants, especially those born on Bikini seeking to go back to their ancestral home, had been denied return.

The problem, according to Leo, lay in minute traces of two radioactive isotopes: Strontium 90 and Cesium 137. Chemically indistinguishable from the natural non-radioactive versions of these elements, they wound up in coconuts and other edible staples of the Islanders' diet.

Despite incredibly small quantities, Leo opined that the offending isotopes would barely fill a teaspoon if the technology existed to separate them from their wholesome brethren, yet it was enough to deny the return of the exiled natives.

Our business in the Marshalls, after the usual meetings with local officials, was a visit to an Island called Wotho, which had recently been visited by a typhoon. To get to Wotho, we took a Sunday morning shuttle flight from Majuro to Kwajalein, a large atoll and home to a major U.S. Navy base.

From Kwajalein, we chartered a 'puddle jumper' to Wotho, met the local dignitaries for a quick tour of the island, then took off back to Kwajalein for the shuttle back to Majuro, arriving that same Sunday evening. All of which took 5 days . . .

Let me explain.

The Marshalls lie west of the International Date Line, but Kwajalein – being a U.S. base – uses U.S. dates. So we left Majuro on Sunday morning (Marshallese date), arriving an hour or so later in Kwajalein on Saturday (U.S.). Around noon, we arrived on Wotho on Sunday (Marshall's date), then back to Kwajalein on Saturday and on to Majuro arriving on Sunday evening.

Since there was no provision in the official work form for such to and fro-ing over the International Date line, by accurately logging date and time, we accomplished 5 days work in a single day! Go figure.

There were a couple of other things that made the Marshalls special, apart from atom bombs and the date line. One of these was Stick Charts, a vanishing skill that survives only in the Stick Charts sold to tourists.

A STICK CHART
traditional ocean navigation device

In the vastness of the Pacific Ocean, atolls are widely dispersed and many days apart for primitive sailors in outrigger canoes. Stick Charts represent the distortion of the ocean swells by the seamounts on which lie the atoll islands, a phenomenon known as diffraction in wave theory. Cowrie shells represent the Atolls on the Stick Charts.

Functional or not, they are beautiful works of art.

There were numerous stories of skeptical U.S. naval officers during WWII who had challenged the skills of local navigators. My favorite involved a Marshallese 'navigator' confined below deck, while the skipper steered his ship erratically around for a day or two, before asking directions to various Islands. The ancient navigator, with only the rolling of the ship in the ocean swells for information, was able to correctly direct the ship to the target island, and several other alternates: a traditional talent sadly lost to the compass and GPS.

The other piece of information that I retained from my short sojourn in the Marshalls involved Arno Atoll, which unfortunately, or maybe fortunately, we didn't visit. Whether to restrict inbreeding, or due to the limited options available to young men among the small populations of their home Islands, there was a brisk inter-island trade of the matrimonial kind.

Nubile young ladies, rightly or wrongly, commanded a price – the dowry thing – and by far the most desirable, and expensive, were the ladies of Arno Atoll. Why I cannot say. That is for your speculation, dear reader. However there was talk about young Arno girls being trained to perform pelvic gyrations on coconuts . . .

I did send a postcard to my dear friends Dave, John and Jimmy Jensen, at Jensen's Marina on Captiva Island – all three eligible bachelors at the time – suggesting that they might entertain a visit to Arno Atoll.

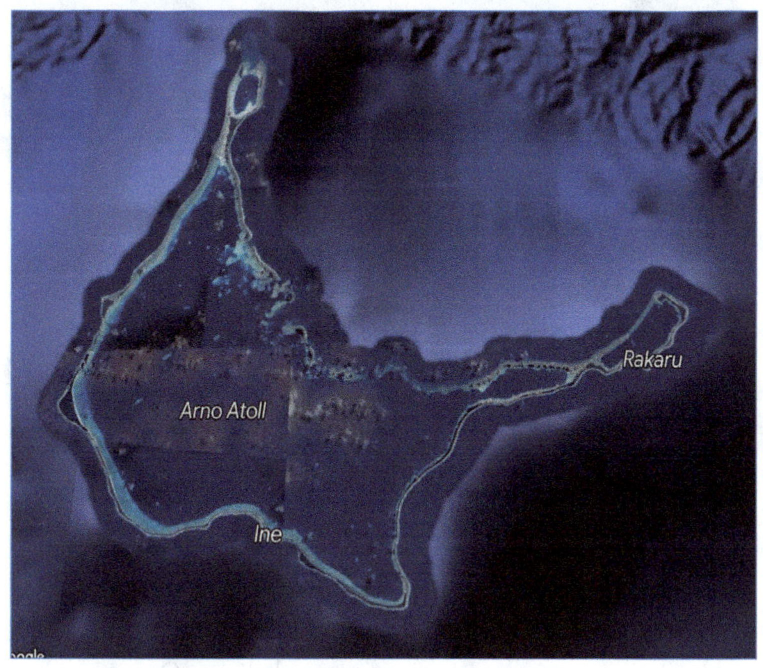

ARNO ATOLL

TRUK AND THE HALL ISLANDS

Micronesia 1990-91

Hopping on the weekly Continental Airlines shuttle from Honolulu to Guam, we left Majuro for Truk, another archipelago. Its capital and most of the population reside on Weno Island in Truk lagoon.

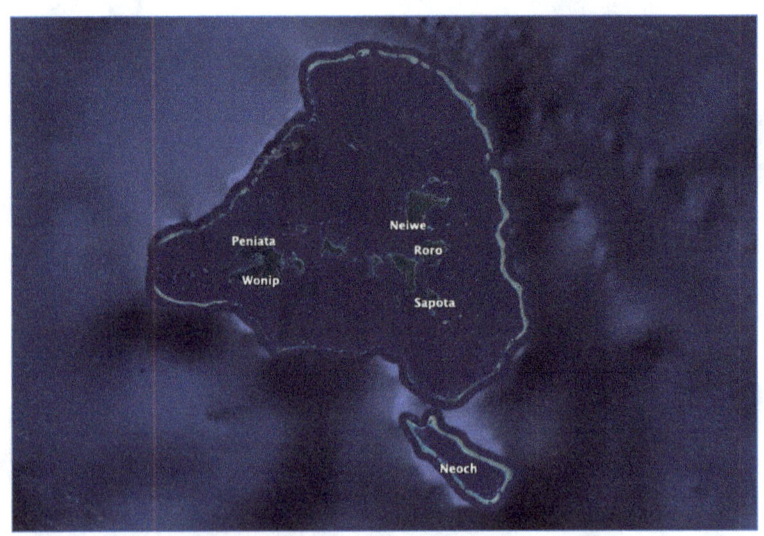

TRUK (NOW CHUUK) ATOLL
a wreck divers paradise

A Japanese hub during WWII, Truk lagoon is famous for its wrecks, and has built a tourist business around wreck diving. Over a hundred ships lie on its lagoon floor: warships and freighters, mostly the latter, many still with their cargoes on board. They were sunk on February 17th and 18th, 1944, in a surprise attack by US aircraft often referred to as "Pearl Harbor's revenge."

After almost half a century, all manner of corals, assorted marine growth, schools of fish and crustaceans had made the wrecks home, creating an eerie duality between a dead ship and a living reef . . . a sort of "now you see me, now you don't." An environmental resurrection of sorts.

Needless to say, wreck diving was not part of typhoon mitigation studies, but nonetheless couldn't be passed up. So we did some typhoon mitigation work to justify the vast expense that Uncle Sam's FEMA lavished on the mission. Not us, the humble functionaries, I hasten to add . . . so we met with Mayors, Fire Chiefs and first responders between wreck dives, most of whom just needed more money.

We also made a side trip to some out-islands called the Hall Islands, which – like Wotho in the Marshalls – had recently been visited by a typhoon, a big one this time.

THE GOOD SHIP MICRO DAWN

The trip to the Hall Islands required a couple of days on a tramp steamer called the Micro Dawn, which still had small patches of paint scattered in large areas of rust . . . like islands in the ocean. Forgive me if I exaggerate: it's more fun. But it was a pretty scruffy vessel.

However, the crew were a lot of fun and mostly high on some narcotic chew like Beetel, but somehow also managed to keep the old ship afloat; avoid the reefs; and deliver us to our destination, through some blended talent of narcotics and navigation unique to these seafaring people.

On board, we had emergency supplies for the devastated Island inhabitants: cases of canned mackerel in tomato sauce, and all kinds of essentials, from matches to toilet paper. All had been carefully calibrated to meet universal nutrition and survival requirements for the destitute, whether Ethiopian or Micronesian.

FEMA EMERGENCY RATIONS
for the Hall Islands.

On board were other provisions that had nothing to do with us or FEMA, and may have been charitable, but were more likely commercial, from members of the crew.

They were giant Turtles, on their backs and tied up, lest they manage to turn themselves over and escape overboard; kept fresh by being kept alive . . .

Local/Commercial Emergency Supplies

Upon our arrival in the Hall Island lagoon, we were greeted by islanders in their canoes or outboard-powered boats, with gifts of fresh fish, coconuts and other delicacies, which we swapped for cases of canned mackerel in tomato sauce.

Once on island, the gifted cases of canned mackerel proved a problem without a can opener. The problem was eventually solved with a blunt instrument and a rock to expose the cans' contents. But neither the appearance nor the taste of the contents was acceptable to the locals and the contents were unceremoniously cast aside, including the cans. Hall Island's pigs, however, were not so squeamish as the locals, and happily ate the contents, leaving the cans to start a non-biodegradable garbage dump.

It didn't take long to realize that the Islanders were way ahead of us. For generations, they had adapted to the reality of typhoons. They built their houses with steep A-frames, which do not create aerofoils to lift shallow pitch roofs in high winds. Ok, the palm fronds get beat up, but they shed the rain and lift when there is a pressure differential between inside and out, and are easily replaced after the storm.

'A-FRAMES'
environmentally appropriate structures

Furthermore, their structures were anchored on posts and palm trees, about 50/50, and lashed together with fiber ropes that shrink and tighten when wet from storms. Better still, they don't weaken like metal hurricane straps when subjected to repeated buffeting.

The reefs absorb a typhoon's wave energy but not its storm surge, which is more of a super high tide. Tying boats, canoes or rafts between coconut palms, with the vulnerable on board, solves that problem. Like a tethered version of Noah's Ark, the boats float above the surge, and return to earth where they started, instead of on Mount Ararat.

The benign tropical climate eliminates the need for blankets and such, and food is not an issue, with an ocean full of fish that know how to survive typhoons. Washed-up coconuts and undisturbed root crops like cassava make up for the dietary loss of a few pigs and chickens, so the only thing we could find to justify our mission to the Hall Islands was protecting potable water.

Typically, coral islands are porous. Rainwater percolates down through the coral sand to a tidal saltwater table. In the porous rock and sand, there is minimal mixing, so the lighter freshwater accumulates, depressing the denser saltwater table to form a 'lens' several feet deep and easily accessed by a shallow well.

All that was needed was a good way to seal the island water well(s) from saltwater intrusion - much cheaper than delivering cases of unwanted canned mackerel in tomato sauce.

Pohnpei Island

1990-91

Our next stop on the Continental shuttle was the Island of Pohnpei.

An extinct volcano rising almost 3,000 feet above the ocean, Pohnpei was very different from the low lying atolls of most of Micronesia. Like the Hawaiian islands, its elevation generates heavy rainfall from moisture-laden prevailing winds, averaging some 700 inches annually, and as much as 1200 inches in the central highlands.

To use Leo's colorful description, "It rains like a cow pissing on a flat rock."

Pohnpei is famous for its coffee, which thrives on volcanic soil in warm wet climates, like its better known competitor Hawaii. Pohnpei's other claim to fame is Nan Madol, massive stone structures built a thousand years ago by a mysterious ancient civilization which collapsed in the 17th century.

More recent ruins include gun emplacements and other fortifications from the Japanese occupation in World War II.

Here again we were able to insert some typhoon mitigation work into our busy sightseeing schedule . . .

Pohnpei Island

Yap Island (now Waab)

Micronesia 1990-91

Back on the Continental shuttle our next stop was Yap Island.

Yap was a real jewel, clinging with determination to its traditions and values. It was known as "The Island of Stone Money," for the giant stone wheels parked around its villages. The 'stones,' some as much as six feet in diameter, are set upright in the ground and just deep enough to be stable. Like flat donuts, each had a sculpted hole in the center and were – perhaps still are – quite literally, the island's money.

Stone money, Yap Island

Rocks and rocks of cash!

Ownership of the stones, which were too heavy to move easily, was transferred ritually, in situ, in the presence of witnesses for the agreed-upon transaction: some land, a boat, a house, a wife, or whatever, who knows. The stone stays where it is, but with a new owner – affirmed, witnessed and accepted by all – this might be the origin of collateral, perhaps?

Each stone has a unique history, its value determined by its story. The stones were plundered from the Island of Palau, now a short 280-mile hop in an Airbus, but not in a paddle-and-sail-powered war canoe.

How or why the folks on Palau were making the stones, no one seemed to know. Perhaps they were pioneers of Stone Age wheel technology. The stones that made it – never mind those lost in transit – arrived with legendary value. How many battles were fought, how many men died in the raid, let alone the journey to and from Palau.

The practice ended, I'm told, when an entrepreneurial Irishman back in the 1800s decided to get rich quick by manufacturing stone money (notice I didn't say printing stone money) and shipping it to Yap, first inflating and then collapsing the stone money economy.

Apart from stone money, the islanders had maintained their traditional values, resisting the temptation to adopt 'foreign' ways. Their villages were traditional A-frame palm-thatched dwellings around a central common area. The men wore loincloths and the women grass skirts, considering it indecent to show their thighs in public. Above the waist was fine, however, so the ladies all went topless.

Welcome dance in Yap
(now called Waab)

The Chief, somewhat of a celebrity, with his photo in the international guidebook, spent most of the day squatting naked, but for his loincloth, preparing his narcotic chew in a pestle and mortar.

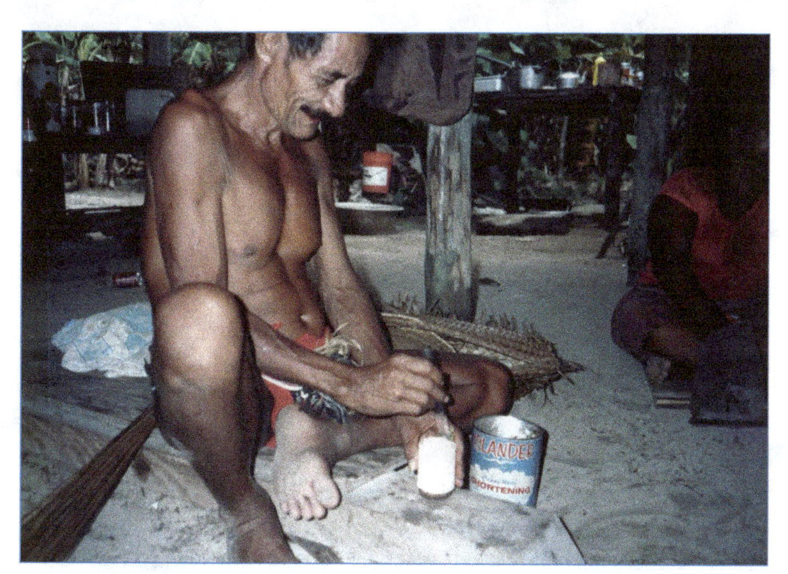

The Island Chief preparing his chew

The ingredients included a fibrous narcotic root; some crushed limestone; betel; and other special ingredients which he ground meticulously in his bowl into a fibrous paste, and then packed hamster-like into his cheek. The effectiveness of his chew lasted little more than the time it took to prepare the next chew, but he was a delightful source of Island history and entertainment while preparing his next fix.

The Chief also gave me a lovely giant mussel shell with the tapered end beautifully wrapped in a woven coconut fiber to make a handle. When I asked him what it was for, he said it was 'wife money,' the traditional trade for a wife . . . ! (Some weeks later, when I got home I showed it to my wife Loretta, who promptly confiscated it . . .)

THE CHIEF PRESENTING ME WITH 'WIFE MONEY'

As usual, somehow we found time for some underwater recreation after our brief 'busy' work day. The exceptional dive experience in Yap is Manta Ray bay. Across the narrow inlet to the bay is a coral reef that acts as a barrier to the tide flow. The incoming tide races over the top of the reef into the bay, sweeping away lots of unwary reef creatures with it, a veritable smorgasbord for the marine filter-feeding class.

And guess who shows up, like senior citizens at an early bird special, Manta Rays, whole schools of them. Like giant B2 bombers gently flapping their twelve-foot delta wings to maintain station, they just sit there sucking up the nutrient-rich flow across the reef.

Meanwhile, just a couple of feet below them, sheltered from the current by the reef wall, divers can literally reach up and scratch their bellies. No kidding!

PALAU (NOW BELOU)

Micronesia 1990-91

Next stop on the Continental shuttle was Palau. Another beautiful Pacific Island, Palau is a little under 280 miles southeast of Yap, and mostly interesting for its local history, but largely overwhelmed by recent history as a Japanese military colony in World War II.

It's also famous for the Blue Corner, a diving Mecca for those seeking shark encounters. A tidal phenomenon in crystal clear water, the Blue Corner is a nutrient-rich environment which attracts every member of the marine food chain, especially sharks! For the thrill-seeker who wants to mingle with schools of large Pelagic sharks, there is probably no better place.

Another Japanese base in WW2, Palau was littered with the remnants of fortifications and some interesting relics. Among these was a World War II Japanese Zero fighter that had ditched on a shallow reef but was still remarkably intact. I was able to pose in the cockpit but couldn't qualify for an entry in my log book.

IN THE COCKPIT OF A REMARKABLY INTACT WWII JAPANESE ZERO

The Marianas:
Guam, Saipan, Tinian and Rota

Micronesia 1990-91

Next stop was the Marianas, a string of seamounts, volcanic islands west of The Marianas Trench, the deepest feature on planet Earth and home of the Challenger Deep, seven miles down.

For geologists, the mystery of such a deep trench was finally resolved in the late twentieth century with one of the last pieces (so far) of tectonic plate theory. It turns out that the fifteen-hundred-mile-long trench is the subduction zone that consumes, after many millions of years, the new ocean floor spreading out from the upwelling magma of the Pacific ridge, eight thousand miles to the east.

Having gathered organic detritus for a hundred and seventy million years, the west-bound seabed of the Pacific plate creeps along at the speed of a growing fingernail, relentlessly gaining weight (don't we all) from the endless rain of the ocean's debris, and finally sinks below the Philippine plate into the red-hot subcrustal magma, to be melted down for recycling.

After cooking for an appropriate few million years, the accumulated organic waste reverts to its constituents: carbon dioxide, water and a few other gasses, which dissolve in the subterranean magma under unimaginable pressure. Such pressure lifts the earth's crust, which cracks, creating fissures through which the pent-up magma can periodically escape.

These events, called volcanoes, return the heat-treated former sea bed to the surface and the atmosphere in an endless global recycling process. If the magma is 'sticky' (viscous), the dissolved gasses erupt explosively, like popping a Champagne cork, only bigger, a lot bigger... Santorini, Vesuvius, Krakatoa, Aetna, Mount St. Helens and Pinatubo, for example.

However, most –70% they say – occur in the deep ocean, where the earth's crust is thinner. And they went unnoticed until discovered in the late '70s by Bob Ballard in his deep-diving Alvin submersible. These 'black smokers' – deep ocean volcanic vents – first confirmed the existence of ongoing sub-marine volcanic activity.

Since then many more instances have been found: mud volcanoes, submarine lava flows, deep ocean vents belching superheated water jets loaded with dissolved gasses and solids spewing plumes into the marine environment just as their land-based volcanoes do into the atmosphere.

So what does volcanism have to do with Typhoons, one may well ask. Well, actually quite a lot. There are, as yet, poorly understood cycles in both the Atlantic and Pacific tropical seas, called oscillations, but also sometimes called decadal oscillations, when the period is counted in decades. These are associated with temperature changes in the ocean surface: warmer water spawns more tropical storms, cold water, good fishing.

Among the original and best known oscillations is the Eastern Pacific El Niño / El Niña oscillation off the coast of Chile. The El Niño, a fishing bonanza, occurs periodically in late December, associated with an upwelling of cold deep ocean water, bringing with it nutrient-rich water to the surface, where sunlight triggers an explosion of phytoplankton, the feedstock of the ocean food chain. These blooms can be seen from space, and – miraculously – by every creature in the oceans, from the microscopic zooplankton to the giant blue whales... all of whom find them without help from GPS!

The warm cycle, El Niña, triggers typhoons, not fish.

As yet, there are no numbers for the enormous amount of geothermal heat energy released into the frigid ocean deep . . . explaining, perhaps, why the burden of responsibility for global warming is laid on the back of fossil fuels.

But back to the immediate past, the Mariana island chain was the site of brutal conflicts in World War II, as U.S. Marines wrested the islands one by one from the Japanese. Guam, the largest and southernmost of the Marianas, remains a U.S. Territory and military base, which dominates the Island. Guam's main claim to fame, besides the military, is a plague of brown snakes which arrived somehow after WWII and took over the Island. Adept tree climbers, the snakes seriously upended the indigenous fauna, especially birds.

Guam also hosts NOAA's Pacific Observatory, with full color lookdown capability from satellites in real time. Covering the whole Pacific, with the ability to zoom in to any area of interest for typhoon formation, was amazing at the time and must be even more so now, thirty years later.

While we were there, Gulf War One started with the air war prelude to the liberation of Kuwait, an event which provided a special thrill for beachgoers off Pati Point.

Andersen Air Force base, in the NE corner of Guam, was home to a squadron or two of the giant 8-engine B52 bombers.

Fully, if not overloaded, these giant machines, engines wide open, boosted with screaming JATOs* and no concern for thrust reducing noise suppression, lumbered down the runway to Pati Point, where a low cliff separated land from the Pacific Ocean. Next stop after a mid-air refuel was Iraq.

Sitting on the beach right below the end of the runway, as one after another of these massive machines lumbered skyward barely a hundred feet overhead, was as dreadful and exciting as it was deafening.

Just north of Guam is the next seamount Island of Rota to which we made a lightning one day visit, other than which I remember nothing and after which it was time to go to Saipan.

Guam

*JATO: Jet Assisted Take Off . . . typically 3,000-pound thrust rockets that burn for a minute or so to bring the plane up to flying speed before being jettisoned (and they are LOUD!)

Saipan, Rota and Tinian

Micronesia 1990-91

From Guam, we shuttled around on ferries and puddle jumpers to the other Islands of the Marianas: Saipan, Rota, and Tinian.

Saipan was another major battle ground of WWII, with wrecks of landing craft still stuck on reefs. There was also a notorious shrine where the cornered remnants of Japanese civilians had hurled themselves off the Banzai Cliffs – with their children – to be dashed to pieces on the jagged rocks below, rather than surrender.

It was impossible to stand there at the Shinto Shrine on the Banzai cliff, side by side with Japanese tourists, and not be humbled by the ghosts and sacrifices – on both sides – by our antecedent generations.

Less than half a century later, Saipan had become a favorite Japanese tourist resort characterized by shooting galleries, which dominated the recreational activities offered on Main Street, offering everything from AKs to burp guns! Only karaoke bars seemed to compete with shooting galleries for the Japanese Yen from a nation that had renounced militarism after Nagasaki.

Just South of Saipan is the Island of Tinian, from where the long-range B29 Bombers launched their raids on Japan in 1945, including the memorialized bomb pits from which the atomic bombs for Hiroshima and Nagasaki were loaded into the belly of the Enola Gay.

TINIAN BOMB PIT

CHAPTER 7:
A Storm in the Desert, Kuwait and Iraq (1991)

KUWAIT AND THE HONEYMOON HOTEL

Kuwait 1991

Kuwait was a mess. The Iraqi army had left in a stampede after looting anything and everything of value they could carry, and setting fire to everything else.

KUWAIT CITY,
the road to Baghdad, Iraq

We had arrived in Dhahran, Saudi Arabia, the vast petroleum complex (formerly Aramco), that fueled the world and filled the Saudi's bank account (and still does). There we got issued some kit, attended a few briefings and spent the night, before setting off in the morning for the few hours drive down the road to the Kuwaiti border.

About half way down the road, we came to the little resort town of Khafji, a favorite beach resort for Saudi honeymooners, and which had recently been the site of a battle following a surprise incursion by an Iraqi column.

Apart from a bunch of bullet pock-marks, there wasn't much battle damage other than a huge hole in the front of the only substantial building, the Honeymoon Hotel itself, where the Iraqis had set up their headquarters after seizing the town.

Apparently the Allied C-in-C, General Schwartzkoff – after determining that it was a fairly small force – decided that this would be a good occasion for our allies, the Saudi and Emirate forces in the sector, to show their mettle.

According to my source, the Saudi / Emirati forces responded with enthusiasm to the challenge, but had a problem locating the quartermaster holding the key to the armory, to requisition the necessary "big stuff," when small arms fire had failed to get the Iraqi invaders' attention.

Among the Royal Rulers of the region, dividing the military into separate commands is a common practice, and for good reason: it makes it much harder to stage a military coup. If there are two armies – say, a National Defense Force and a Palace Guard – under separate commands, and one has the tanks and the other the anti-tank weapons, it allows rulers to sleep better at night (and has, so far, dodged the fate of most of the former Middle Eastern monarchies, from Libya to Pakistan, that were outed by Military coups).

So it came to pass, as I was told, that the Iraqi commander, holed up in the Honeymoon Hotel and unimpressed with small-arms fire, declined every invitation to surrender. Unable to locate the quartermaster – the one with the key to the store where they kept the big guns – the battle stalled.

It wasn't until a nearby unit of U.S. Marines provided a TOW missile

– a powerful wire-guided anti-tank weapon – and blew the huge hole in the front of the hotel, that the Iraqi commander changed his mind . . . and waving a clean white sheet (courtesy of the hotel housekeeping department), surrendered!

Soon after Kafjie, we reached the Kuwaiti border where we were welcomed like conquering heroes, which was nice but a bit embarrassing, since the cease-fire had been called a few days prior to our arrival, and I suspect had more to do with the Third Armor, the 82nd Airborne and the SAS than with the arrival of we three. Nevertheless, somehow, we managed to find our hotel and check in.

The retreating Iraqi army had trashed and burned everything that they could not loot, so habitable accommodations were scarce. Our hotel had also been trashed and burned, except for the one wing in which we were lodged. Our fellow guests were the firm of Boots & Coots, the oil well firefighters made famous by Red Adair.

There wasn't much to do of an evening in Kuwait except get a bite to eat in one of the local sidewalk 'restaurants,' and then find somewhere for a drink. Although officially banned, alcohol was not hard to come by, but one had to know where to go.

Before the war, for example, the sky top restaurant in the Intercontinental hotel was booze-friendly to western infidels but not so to eastern faithfuls from Pakistan or Indonesia. Nevertheless one had to know the protocol.

To the waiter's opening solicitation "what may I serve you?" the response was always 'chai' (tea).

It was the follow-up question "what flavor would you prefer" that was key. Answer "Johnny Walker" and – hey! presto! – teacups and a teapot full of Johnny Walker whisky would arrive.

The Iraqis had no such issues with alcohol, and after six months of occupation the Kuwaitis had abandoned any and all pretense at Islamic prohibition. One of our preferred haunts was the SAS Hotel, which unlike most of Kuwait's hotel had not been totally destroyed. It was a bunch of little beach-front cottages, with a main building containing a bar and restaurant around a swimming pool. Happily, the bar was fully operational.

When I say it was on the beach, it was. It was not a particularly beautiful beach, as beaches go, especially if you've seen beaches in Pensacola or the Bahamas and plenty of other beautiful white sand / blue water stretches. But in the buildup to the invasion the Iraqis had been led to expect a traditional WWII U.S. Marine assault on the beach, Iwo Jima style, so they had heavily defended the beach.

Apparently, no one had advised them of the radical change in tactics since Iwo Jima, initiated, they say, by Col. John Boyd who asked, "Why attack the beach? Let's attack the Enemy!"

IRAQI DUGOUT
on the beach at the SAS hotel

The Marines were a feint: a foil for the real assault, which was a left hook through the desert to cut off the elite Iraqi Republican Guard in Safwan and Basra. As General George S. Patton once described the tactic: "Grab 'em by the nose, and kick 'em in the ass! "

As a functioning bar, the SAS Hotel was a waterhole to which we repaired frequently. The principal occupants of the Hotel was UNIFIL, the UN force charged with monitoring the ceasefire while the Allies disengaged. The force was made up of contingents from Sri Lanka, Fiji, and other nations which lacked a common language (and besides which, they were waiting for the arrival of their brand new blue and white Land Rovers with UN insignia before going to work).

Needless to say, the Iraqis had mined the beach and set up all manner of machine gun nests and boobytraps to discourage marines who wisely never showed up.

To return it to its proper function as an asset for hotel guests, a local contractor had been hired to clean up the beach, and had parked his sizable tractor-trailer in the empty hotel parking lot, using it as the repository of all the unexploded ordnance recovered from the beach. About the time that the shiny brand new U.N. Land Rovers arrived, and were neatly parked next to the contractor's vehicle in the otherwise empty hotel parking lot, strangely, the contractor's truck caught fire.

Soon the ordinance started to explode. Not all at once, but with increasing intensity, a sort of Tchaikovsky's 1812 crescendo of tracers, grenades and RPGs . . . an orchestra of bedlam, punctuated by the big boom tympani of an anti-tank mine, or a mortar shattering the evening tranquility.

The UNIFIL forces, united by the single word they shared, raced around the hotel grounds diving under tables shouting "ATTACK! ATTACK!" as they sought safety behind hedges or in the empty swimming pool . . . all the while shouting the one word they all understood: ATTACK!

The next morning, all that remained of the half dozen or so brand new Land Rovers was a number of silver-gray puddles of frozen aluminum in the parking lot! None of which, as I happily recall, interrupted the essential service of the bar.

Meanwhile there were 980 oil wells blazing away in the Burgan and other oil fields around the city, and our roommates from Boots and Coots were still scratching their heads about where to start, let alone what to do about them.

BURNING OIL WELLS EVERYWHERE

So we frequently repaired to the place where we could sip away their sorrows.

They explained that their expertise was in dealing with single blowout wells. Their standard procedure was to blow out the flame with dynamite, quickly inspect and measure the still-gushing wellhead, all the while being hosed down by a colleague with a firehose and hoping the well didn't reignite. Retreating with data in hand, they would reignite the still-gushing well and return with a suitably modified well cap to install and shut down the well.

This was particularly relevant in Kuwait, because, unlike the typical 'wildcats,' where ultra-high pressure gas pockets miles down surface blow out the drill pipe, leaving the surrounding well casement intact, the Kuwaiti wells were producing wells that had been deliberately destroyed with explosives, so the casement housing the wells was no longer round and easy to fit.
In addition, there were hundreds of burning wells in every direction. So the prospect of a flash back as the oil-gas plume hit the next flame compounded the danger and the urgency. Consequently they were nibbling away at the perimeter wherever they felt they were downwind of the inferno for a few hours.

Around many of the wells, there were lakes of burning tar which grew larger by the day as the gas and volatile ingredients of the plume burned and the 'heavy' residue stuff rained down in drops of liquid tar. Needless to say, this complicated access to many of the wells.

There was a story going around about a film crew from a major news agency that had driven their leased 'all terrain' four wheel drive vehicle into one of these lakes. Apparently 'all terrain' did not include burning tar sand! The tires melted and caught fire and the engine stalled in the oxygen depleted air with disastrous consequences for both vehicle and its occupants.

Everywhere above the fires there hung a great black cloud that turned high noon into twilight and rained down a sort of drizzle of tiny tar drops. Vehicles acquired a speckled look, as did clothing, especially whites. Laundry required multiple doses of industrial strength detergent but shampoo was totally ineffective for anyone foolish enough to go without a hat!

Apparently a team of scientists from EPA, charged with studying the long term impact of such unprecedented pollution, quickly relocated to Ras Tanura, 300 miles down the road, to study it from a safe distance.

At the time, which was only a few weeks after the ceasefire, Boots & Coots had an exclusive with the Kuwaiti government, and were predicting years, not months, to get the job done. So the Kuwaiti Government, watching the nation's assets go up in smoke, voided the contract and wisely threw it open to the unqualified: the hoi polloi, as the elites of Ancient Greece described the common folk.

The new deal was simple: pick a well, put it out and we'll give you $10 million. Period!

Well, guess what, in the ensuing unregulated frenzy of trial and error, Poland's Warsaw Fire Brigade apparently came up trumps!
An old Russian jet engine on a flatbed truck, fired up with a fire hose feeding the intake and backed up to the flaming well head, snuffed out the flame with an oxygen-depleted exhaust of steam and carbon dioxide, allowing access to the well head to cap the well . . . kind of like blowing out the candles on your birthday cake, only you got multiple puffs if the first one didn't work, or the plume reignited. Better still, the jet's exhaust diluted the oil gas plume enough to reduce the risk of re-ignition or flashback from a nearby burning well.

In less than three months the fires were out and Kuwait's bank account secured, if a little depleted.

Far less dramatic, our mission in all this was a liaison function. This involved much to-ing and fro-ing between Safwan in Southern Iraq where the ceasefire line and the action was and Kuwait City where the US Embassy, our HQ, was located.

In Safwan, the three of us – two guys and a gal – were billeted in a fifty-man tent (actually 48 men and two women) with a Civil Affairs battalion (C.A for short) populated almost entirely with Mormons, an epiphany for me. Apparently the U.S. Military depends heavily on Mormons as foreign linguists, because of the missionary obligation of their faith.

The ribald comments of the waking soldiers were neither pious nor in any way respectful of the two women in the tent . . . another epiphany! The battalion was attached to the Third Armored Division under the command of a two-star General to whom we reported. Like other cavalry officers of my acquaintance, he enjoyed wearing jodhpurs, long after battle horses were obsolete . . . and carried a riding crop with which to punctuate his commands, by smacking his riding boot.

We were provided with a couple of Humvees in which to rush around, which was fun, despite the fact that being military, they were designed for abuse not comfort. Twice a week, one of us would drive to Kuwait City, down the aptly-named Highway to Hell, which ran smack dab through the hundreds of burning oil wells in the Burgan oil field.

THE APTLY-NAMED HIGHWAY TO HELL

In Kuwait, we would stop by the U.S. Embassy – where we had a sort of back door key – brief the Ambassador on what was going on in Safwan, then check back into the unburned wing of our hotel with the Boots and Coots chaps, to share their woes over some 'liberated' beverage or other. Everything in Kuwait had been 'liberated' in those days.

Then, after rushing around Kuwait City in the Humvee for three days, it was time to go back to Safwan, to brief the General on what was happening in Kuwait.

Iraq Riots and Refugees

Safwan, Iraq; Gulf War One 1991

Among the strange derivations that evolve from noble beginnings like the Geneva Conventions, designed to make warfare nice – like the right way to kill your enemies and how to treat prisoners etc. – came the need for referees to make sure that the rules are followed, as in boxing or wrestling.

Referees must be scrupulously neutral, which means combatants cannot be trusted to deal directly with each other, so intermediaries are used. At the upper levels, this is called shuttle diplomacy. At our low level, it's called liaison.

The issue in Iraq was refugees: a tidal wave of people pouring through the cease-fire lines, trying to get out of Iraq.

The Kuwaitis didn't want them, the Saudis wouldn't take them, and they refused to go back to Iraq while Saddam was in charge – and for good reason – so they were trapped behind the Allied lines, but still on Iraqi soil.

This meant that the UN High Commission for Refugees (UNHCR) declined to accept responsibility for them, since they had not crossed the border, essential to qualify as a refuge . . . thus leaving them in the tender care of General Schwarzkopf and his coalition forces who, as signatories to the Geneva Accords, were obliged to provide for their safety and welfare and – worse still – could not go home until someone relieved them of their responsibilities.

That's where we came in.

Some neutral entity had to be contracted to relieve General Schwartzkoff of his obligation, by finding countries outside Iraq that would take the refugees. So we, two laddies and a lass, wound up taking over the 'Refugee camp' (which was actually a displaced person camp) of some 20,000 souls desperate to escape Saddam's Iraq, but still confined to Iraqi sovereign territory occupied by the coalition forces. Easier said than done.

The military had set up the camp on a grid system with Avenues A, B, C intersected with streets 1, 2, 3 etc., sort of like Manhattan, except surrounded with barbed wire.

Each arrival was searched, allocated a plot like: Ave. B, street 3, number 4, issued a tent and a care package, and – bingo! – on to the next one. All of which makes sense unless, of course, you're in Iraq, where everybody was justifiably paranoid. At the time, there were reputed to be 23 secret agencies all reporting directly to Saddam. By some estimates, one in five of the population worked for one of these secret spy agencies . . . so nobody trusted anybody, not even family. And the grid system didn't work. Not knowing who your neighbor was, in the tent 10 feet away, just fed the paranoia.

Weapons were smuggled in, and fights were endemic.

One night, for example, we were woken up at 2 am to deal with a riot in the camp. A truck had crashed into the fence and the driver was being mercilessly beaten up by the inmates. By the time we arrived, which was soon, the truck driver had been beaten senseless, or so it seemed. We managed to extract him from the vengeful mob and get him to a safe place to repair his damage and interrogate him.

It turned out that his confusion was more alcohol-related than from the beating by the mob. He had actually fallen asleep at the wheel, and driven off the road, there being little difference as far as bumpiness between a bad road and the bordering desert. His unguided vehicle had wandered off-road until it encountered the camp fence. A message, perhaps, for those who believe in self-driving vehicles. The inhabitants had assumed that he was an assassin sent by Saddam, and reacted accordingly.

Working in this environment was a bit like sitting on a powder keg while smoking a cigar!

On another occasion, I had driven into downtown Safwan in our coalition-provided Humvee, which wore a big black upside down V (the Arabic numeral eight), to identify friend from foe . . . when I noticed a very beat-up old Chevy driving slowly down the main drag just ahead of me. Something about his behavior set off my 'trouble' antenna. He seemed to be stalking someone or casing something, but clearly was up to no good.

Curious, I hung back to watch. I remember noticing a UN vehicle, probably a UNIFIL observer, parked across the road at the end of the street, perhaps a couple of hundred yards ahead.

Before long, my suspicions were vindicated.

While the car was still rolling, its doors suddenly opened, disgorging several men armed with clubs who attacked a man walking down the sidewalk, in what looked very much like a kidnapping. After a brief skirmish, the victim broke free and took off in the direction of the UN vehicle, gesticulating wildly and shouting the Arabic version of "HELP," hotly pursued by his assailants.

Alarmed by what they must have assumed was a mini riot, a charging mini-mob armed with sticks and led by a screaming maniac, the UN observers fled . . . probably to write up their report in safety.

Incensed by the UN's feckless performance, I hit the horn and the loud pedal full bore, charging from behind with horn blaring and tires squealing, cutting off the stalking Chevy and taking them totally by surprise: always a good thing in attack.

The assailants scattered like hounds and I was able to bring the wild-eyed breathless hare (the victim) on board, and back to base, which, as it happens, was 'our' camp.

This, among other incidents, showed me just how ugly is a society that uses intimidation and violence to control its citizens, rather than one that reflects the norms of ordinary folk, which, although messy at times, are generally pretty benign.

Needless to say, the story went 'round the camp, and though I didn't know it at the time, greatly enhanced our image. It also attracted multiple requests begging for help, from people who believed we had extraordinary power.

Among these was 'a wee brown fella with a mustache,' a description that fit most young Iraqi men, this one approaching me with "Señor, habla Espanol." It was so out of place, that at first I thought I was hallucinating... but it turned out that he was from Mexico, and several years prior had come to Iraq to bury his grandmother. Having some expertise or other, he had been conscripted and sent to the front of the Iran-Iraq war, which he somehow survived, and now wanted to go home!

The camp's internal security problem stemming from the informant-paranoia was solved by a very simple ruse. We reorganized the camp by eliminating the grid and replacing it with neighborhood communities, or villages, Each village was named after an Iraqi town or suburb, and was set up with a community tent in a central square: the town hall if you like, and manned by representatives chosen by the community and with whom we dealt directly.

All we had to do was to search each new arrival for weapons, ask where they were from, and direct them to their community, where credentials were quickly established by former neighbors one way or another, something that was impossible for outsiders to do. Unwanted neighbors and known agents of the Regime were identified and rejected. It was amazing how quickly things settled down once everybody knew who his neighbor was.

At first there were quite a few rejects, who – having no friends in the community – were expelled, with nowhere to go but back from whence they came.

It should come as no surprise that managing twenty thousand people cannot be done by three, even with the backing of the U.S. military. So locals – I won't call them inmates – were recruited to build a hierarchy of volunteers with a variety of necessary skills, including, of course, the ability to interpret languages.

Among the many volunteers, one young man stood out.
His name was Hassan, and his story needs a chapter of its own.

THE GREAT ESCAPE

Kuwait 1991

As often happens in emergencies, exceptional individuals emerge from the mediocracy of the majority. Such folk are the rubber that connects the drive to the road; the yeast, so to speak, in the otherwise unleavened dough of humanity.

One young man stands out among the many desperate people who looked to us for salvation, often literally for their lives. Hassan was a young man born in Iraq and who had lived and worked in Kuwait for a dozen or so years as an account executive in a Kuwaiti advertising agency. He spoke impeccable English and was one of the first in the Iraqi Refugee Camp to come forward and offer to help.

Hassan very quickly became a right-hand man. Intelligent, articulate, charming and hard working, he was an all-round class act. His story, however, was a most harrowing tale of bad luck.

Married, and with two young children, he had gone through the agonizingly long and tedious process of applying for immigration to Canada. After years of paperwork, patience and persistence his efforts were rewarded, his application was approved and his immigration status finally granted.

Packing up his life in Kuwait, he sent his family to the UK to wait while he wrapped up a few loose ends before joining them two weeks later and heading to a new life in Canada.

Four days later Saddam invaded Kuwait.

As an Iraqi national, Hassan was now in double jeopardy: firstly from the Kuwaiti resistance that saw him as an enemy agent, and secondly from the Iraqis who would press him into their service . . . either military, or - worse - intelligence.

Unable to contact his family in the UK, he went back to Iraq to stay with a trusted uncle, where he went into hiding.

In the chaotic aftermath of the war, he was able to cross the ceasefire lines and make it into allied occupied territory, where nine months after sending his family to the UK, he wound up in our camp and approached me to offer his services.

Hassan had been unable to contact his family since the invasion without revealing his location, and as we found out later, they didn't know if he was dead or alive.

Getting the twenty-odd thousand displaced folks to safe haven was key to allowing the coalition forces to go home.

There were actually dozens of International Agencies and NGOs hovering around, as there always are in these situations: ICRC, Save These, Care for Those, you get the picture. But bottom line: the object of the exercise was to find safe haven for the refugees from Saddam's Iraq anywhere but Iraq, where they were stuck, and where none of the neighbor states - Kuwait, Saudi Arabia, UAE etc. - would have anything to do with them.

Anyone born in Iraq was specifically barred from entering Kuwait, even in transit.

And the NGOs were as much a part of the problem as they were part of the solution, clogging up the already broken bureaucratic avenues to immigration.

Iran came up trumps, no pun intended, offering to take in anyone with a relative in Iran, thereby creating a fifth column for the perpetuation of the multi-millennial Persian-Babylonian conflict.

So, thanks to MAC (US Military Airlift Command), we were able to unload over half of our charges, mostly from the bottom of the socioeconomic strata, into the tender care of Iran.

Southern Iraq is mainly Shia, as is Iran, but still a minority behind the Sunnis in the Islamic world. Consequently about half the camp were flown into Iran in the C130s, courtesy of the US military.

LOADING A C130 HERCULES MILITARY STYLE

Most countries, in the western world at least, have quotas for the number of refugees that they will accept annually. So we 'liased' with the NGOs that negotiated with the respective embassies.

Getting Hassan out of his situation and reunited with his family became my obsession.

The first obstacle was the fact that anyone born in Iraq was forbidden by

law from setting foot on Kuwaiti soil, or perhaps I should say Kuwaiti sand . . . and the only way out was through Kuwait.
To make matters worse, the Canadian embassy had been ransacked, and claimed to have no record of his or his family's existence. So they were of no help.

Denied all legitimate avenues, the alternative was obvious: bend or break the rules. Entering Kuwait from Safwan, Iraq was a piece of cake for me: I was doing it several times a week. The exuberance of liberation had not worn off on the Kuwaitis at the border checkpoints, where we were greeted with everything from 'God Bless America' to 'Thank you Mr. Bush' and an assortment of blessings in Arabic.

Between the Humvee and my official pass, getting Hassan into Kuwait was not a problem; finding him a safe house in Kuwait was not impossible but a bit tricky. The real problem was how to get him on a plane to the UK. A civilian airline at the international airport was out of the question: passport identification would have led to his instant arrest. So it had to be a stowaway on a military flight.

Happily, MAC was shuttling multiple flights to Brize Norton, the USAF base in Oxfordshire, UK.

In the early 1960s, as an RAF trainee pilot on this very base, I had learned to speak American - not to be confused with English - while practicing zero visibility ground-controlled approaches (GCAs), being talked down by a ground controller with a deep southern drawl over a distorted radio in a very noisy cockpit.

But I digress. The obvious solution was to smuggle Hassan aboard a MAC transport to Brize.

At the upper levels, such things are simple: a question of who you know. At our humble level, not so much; it has to be done at the operational level where the NCO is supreme.

That meant finding a sympathetic loadmaster of the sergeant variety.

Not a big problem if you know where they hang out for a cold beer or two, which happily we did, and with a little cold beer bribery, a sympathetic loadmaster was soon enlisted to the conspiracy.

Meanwhile, the number one priority was a safe house in which to park Hassan while we sorted out the extraction details. Not so easy. In critical path parlance, that was the 'float time' and had to be as close to the extraction as possible, to minimize the risk of discovery. Private residences were too risky, so the best bet was an embassy.

By International law, embassies are sovereign territory of the home State, not the host State, and so are not subject to the laws of the host State. The US Embassy was sympathetic, but since Hassan was heading to the UK - where his family was - the British Embassy made the most sense.

Before leaving the camp in Iraq, as a safety precaution, Hassan gave me a small package containing all his identification papers: passport, birth certificate, bank accounts, etc. . . . anything that could identify him should something go wrong. I was to return them to him once he was safely in the UK, where I would rendezvous with him on my way back to the USA.

Being entrusted with someone's entire identity documentation is an awesome responsibility, second only to having someone's life in your hands. Any clue to his identity would have gotten him in big trouble with the Kuwaitis, who would likely have roughed him up as a spy before deporting him to Iraq, for even worse treatment.

Now the golden rule when bending the rules states: "It is better to beg forgiveness than to ask permission." And so it was that I was able to drive into the British Embassy compound with Hassan, and my powerful 'go anywhere' pass. That was the easy part. Once inside the compound, I tried to deliver Hassan to the care and safekeeping of the Sovereign territory of the British Embassy, where an apoplectic minor official with a pronounced Pakistani accent ranted and raved in a vain attempt to deny Hassan asylum.

Unapologetically, we prevailed by refusing to leave, and then, moved up the chain of command until we found a sympathetic ear and Hassan was safely installed in the British Embassy, pending the next stage of his escape. Our official with the sympathetic ear also managed to harness the power of Her Majesty's Government to locate Hassan's family and arrange a telephone connection: their first contact in over nine months.

It wasn't long before we got the nod from our conspiratorial loadmaster: a predawn flight without a full complement. It was off to the Embassy, and Hassan was whisked away to the MAC section of the airport and loaded aboard a C141 bound for Brize. Once he was safely airborne, we managed to arrange for his UK-based family to be on the tarmac at Brize as an arrival surprise. All of which worked perfectly.

Some weeks later, on my way home, I stopped in London and contacted Hassan. We arranged to meet at the street-level entrance to the Sloane Square underground. I arrived first, and remember standing on the sidewalk peering down the stairway into the bowels of the London Underground Circle Line and waiting, patiently, and not a little anxiously.

And then he appeared, bounding up the stairs like Orpheus departing the underworld. I shall never forget the joy of that moment, as I handed Hassan his little parcel of documents and we embraced. Rightly or wrongly, getting that fine young man out of his predicament, reunited with his family, and free to pursue his life was more satisfying than all the 20,000 others whom we managed to relocate.

CHAPTER 8:

To Russia with love, and MREs

I Want You to Meet Some of my Friends

Siberia January 1992

“I want you to meet some of my friends,” said the pretty Russian girl who was my interpreter.

“I would like that very much,” I replied without hesitation.

“No! No!” She said lowering her voice “I mean without them” nodding towards the three official ‘tour guides’ who escorted us everywhere. One was a military officer, one was a high-powered political type from Moscow and the third was a senior local official.

In those days one could always tell who was who in Russia by their teeth. Political types had gold teeth, military types had gun-metal teeth and the ordinary folk had no teeth.

Two of our escorts had gold smiles and the biggest one had a steel grin like Oddjob in the James Bond movies.

My interpreter had a very pretty young face peeking out from a thick fur halo that fringed the hood of a bulky Parker and pant suit which made her, like the rest of us look more like the Michelin man in a tire commercial than people.

OUR INTERPRETERS. ULAN UDE 1992
C141 loaded with goodies background

It was January and cold enough to make a penguin shiver.

The temperature fluctuated between minus 20 and minus 60 degrees the whole time we were there. Centigrade or Fahrenheit makes little difference at those temperatures.

We were in Ulan Ude, Outer Mongolia, alleged birthplace of Genghis Khan.

A few short months prior it had been the Soviet Socialist Republic of Buryatia. Now it was part of the the CIS** after the collapse of the USSR and the defection of most of the European and Asian 'Soviet Republics'.

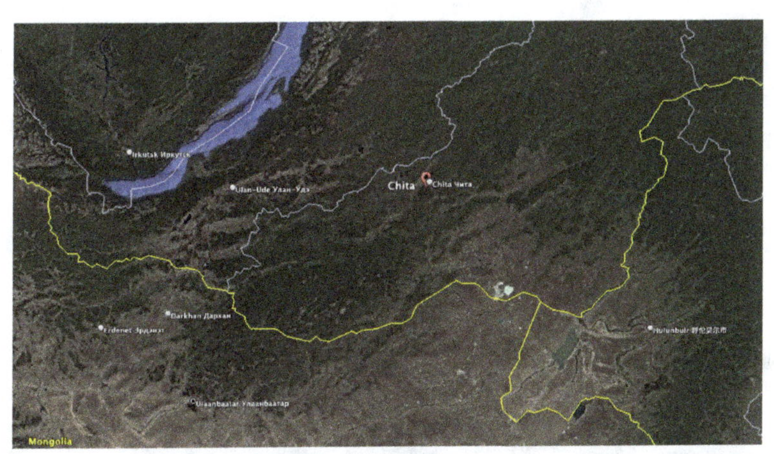

MAP OF THE TRANS BAIKAL AREA OF SIBERIA.

The implosion of the Soviet Union had brought down the central control system which had formerly managed supply and demand however badly. With the collapse of the Union went the central control system, most of the necessities of life were now produced in foreign states with much better markets for their wares. Transbaikal, as the vast area east of Lake Baikal is called, was particularly in trouble.

Richly endowed with brown coal, oil and gas, and littered with mines producing virtually every mineral in Mendeleev's table, the region is sparsely populated. Its towns are centered around mining , manufacturing and military installations and separated by vast empty spaces that produce nothing to eat. It was home to the Soviet 3rd. Army, stationed there in the '60s to keep the Chinese in their place. It was also home to a healthy crop of ICBMs aimed over the North Pole at us, which explains why the region was closed to everyone except Soviet Officials.

When we landed in Ulan Ude in January of 1992 our hosts declared that we were the first 'outsiders' to set foot on the ground for 50 years.

THE FIRST OUTSIDERS IN 50 YEARS

Fifty years prior Hitler's Wehrmacht was at the gates of Moscow and pulverizing Stalingrad.

Now their problem was food.

With a growing season between the spring thaw and autumn freeze of less than four months, agriculture had never been on the Soviet radar for the region.

Happily we had arrived with a giant C141 cargo plane bearing gifts of food, medicine and other essentials courtesy of Uncle Sam; which explains why 4 Americans and a Kiwi were welcomed as the first Outsiders to set foot on the ground in Ulan Ude in half a century.

The day had started as usual with the arrival of a couple of vans containing our escorts and interpreters at the barracks where we were housed. From there we were driven to some government building for the 9 a.m. meeting with the mayor and various department heads of the particular town.

The meetings followed a predictable pattern. Along the center of the conference table around which we convened was a line of green glass torpedo shaped bottles of vodka. These became known as 'artillery'. They were about the size and shape of the business end of an RPG* and to those on the receiving end had a similarly devastating effect.

As each official was introduced he would propose a toast to us and Russian American friendship with a swig of vodka which protocol and common courtesy required us all to follow. As the process moved round the table sooner or later one of our hosts would claim that his father or uncle had been at the river Elbe where the Red Army and the Americans first met in 1945.

This required a 'heavy' toast where the proposer lifts his drinking elbow with his free hand and chugs the whole glass sort of like doing shots. Needless to say protocol and common courtesy required everyone to follow suit. Not wishing to be outdone, every succeeding official had a special friend at the Elbe requiring a 'heavy' toast until the cycle was complete and everyone was trashed.

Then the meeting started.

There followed a question and evasive answer session where we tried to pry the information that each of us had been charged with acquiring. I don't remember much of those meetings but one exchange made its way through the alcoholic haze to my memory.

The State Dept. G12 on the team was going through his list when he asked for some numbers on the homeless population. Our Russian host looked puzzled, turned to the interpreter who bounced the question back. "Homeless, you know, people living on the street" repeated our G12.

Still looking puzzled, the Russian said "None. We don't have any homeless on the streets." Thoroughly skeptical now, our Washington bureaucrat responded: "That's not possible you must have some".

The Russian's expression shifted from puzzled to incredulous. "It is too cold, they would freeze to death," he said.

At 50 below he had a point!

After the meeting we were escorted on a guided tour of institutions that they wanted us to see; a hospital, an orphanage, an old folks home, and, of course, the obligatory Government store so that we could see for ourselves how desperate was their plight and need for aid.

The Government store was typically a large one room storefront with a single stone counter running its whole length on which were plenty of 'artillery', the green glass bottles of Vodka that sold for a few cents a bottle. In the corner of the room was a pile of moldy potatoes, many of them starting to sprout, and a very few overpriced edible items such as a can of soup or a dried fish. Outside, extending for half a block was a line of potato shaped bundled up Russian citizens patiently waiting their turn to see if there was anything other than Vodka and moldy potatoes to spend their money on. Everyone had plenty of rubles because there was nothing to buy.

It was outside the government store when we were wondering how people existed let alone survived in the miserable conditions we had been shown when my interpreter made the proposition.

A plan was hatched to slip our official escorts and that evening, after we had been returned to our quarters, I slipped out at the appointed time for the rendez-vous where my interpreter was waiting with a car. Apart from the official government vehicles that shuttled us around, I don't recall seeing any cars but she found one and a driver to boot who whisked us through the deserted streets to the very drab front door of a nondescript building. The door opened to a coded knock like a prohibition era speak-easy, and we were escorted down a spiral staircase into a large beautifully furnished basement restaurant and seated at a table.

Most of the dozen or so tables were occupied with smartly dressed customers unlike anything we had seen all day. In one corner of the room was a jazz quartet that could have been straight out of New Orleans and around the room hovered immaculately groomed waiters in black ties. The whole scene was so different to anything we had been exposed to during the day that the contrast was quite literally night and day.

A young man who turned out to be the proprietor came over, and after welcoming us in perfect English asked if there was anything in particular I would like. Not convinced that he really did have instant access to pretty much anything and everything, I asked what if any American cigarettes he had.

At the time I was a smoker and had been reduced to the local Russian brands which were foul tasting black tobacco and worse still, dangerous. They consisted of a short densely packed plug of black tobacco on the end of a three inch cardboard tube. The amount of suction needed to draw smoke through the plug could strip the chrome off a trailer hitch, with the result that at some stage the burning tobacco would release and land in the smoker's epiglottis! As a precaution it was necessary to pinch the tube vertically and horizontally to catch, in theory, most of the burning tobacco.

"What brand would you like?" he asked patiently.

Not to be outdone I replied "do you have any Dunhill"?

I was sure he wouldn't have the high end 'designer' brand. Without a blink he replied "red, blue or green" referring to the regular, light or menthol options. It was game, set and match. Whatever he had going, this guy was for real and I wanted in on the story.

After a sumptuous meal, beautifully served, we had a tour of the proprietor's 'underground' operation. The space was pretty limited, but what he had packed into it was amazing.

He had desktop computers, state of the art Wangs or Apple Macintosh (this was just prior to the laptop); by contrast the 'overground' official stuff that we had been shown all day everyday had nothing more sophisticated than a non-functioning NCR mechanical cash register that had been replaced by an abacus, an adding device developed sometime between the Stone age and the Middle Ages.

The various goods and services that he offered went far beyond those of the 'overground' economy, provided of course that you could pay for them.

Of particular note were his big game Hunting Safaris aimed at wealthy trophy hunters mostly from Germany, seeking polar bears, elk, or whatever special trophies the Siberian Arctic offered. Our man had 'arrangements' with senior military officers, for the most part clients of his nightclub, whereby he had access to helicopters, radios, rifles, a fleet of aircraft of all kinds equipped with skis, floats or tundra wheels for every terrain all courtesy of the Russian military. I should add that the ruble at the time was trading on the black market at 100 to the $ versus the official rate of 7 to the $.

To put that in perspective, the 350 rubles domestic air fare from Moscow to Ulan Ude, a distance of over 4,000 miles and 5 time zones was $3.50 at the black market rate versus $50 at the official rate... take your pick. Everything was for sale especially if you had the Almighty $.

The giant AN2 biplane, the workhorse of the region, was available at $20,000.

Thinking that it might afford me some respect in the Bailey's shopping center parking lot on Sanibel, I made a $150 down payment on a brand new Soviet military 'Jeep', a sort of CJ5 on steroids with, as I recall, 13 forward and 3 reverse gears. It had a huge red Hammer and Sickle logo on the hood and a similarly emblazoned Arctic quilt hood cover.

It was on the tarmac when our empty C141 was ready to take us back to Rhein Main, Frankfurt and the load master and I had come to an arrangement that the moment it was offloaded at Rhein Main it would be my responsibility not his, when the pilot, a National Guard reservist, wandered by and got the 'Wobblies'. So we had to leave it on the Tarmac, with my deposit, where no doubt it still is?!

It was not long after our return home that post Soviet Russia had its first real elections. Back home in the States I remember reading the lamentations in the mainstream media about how all the former Communist officials had been returned to power with the inference that rigid totalitarianism was back.

From my experiences in Siberia I suspect that there was another dynamic at play. Under the Soviet system the black market flourished under the patronage of the 'Nomenclatura', the all powerful eight hundred thousand or so elite ruling political class. When the 'free market' was instituted the 'underground economy', freed from the shackles of their political masters, prospered mightily to the chagrin of their former controlling political beneficiaries who now all wanted to be Capitalists too. So they teamed up... the Rulers and the Racketeers.

Trouble was; without rigid central control there were no rules at all, so in no time the market place degenerated into a 'thugly' free-for-all... like Chicago during prohibition.

To make matters worse the Rulers had no entrepreneurial skills and the Racketeers had no regulatory talents so by mutual consent, the bureaucrats ran for office to make some rules and the Racketeers, went back to business as usual switching seamlessly from black market to free market and shedding the stigma of illegality without missing a beat.

I'm sure the mutual advantage, or profit sharing was undisturbed just like old times but with a better balance. Perhaps even with the advantage shifting to the black marketeer, now businessman, as communism morphed into crony capitalism.

A few months earlier I had been commuting twice weekly in a Humvee between Safwan, Iraq and Kuwait city, about a four hour drive along a road known for good reason as the Highway to Hell. The road ran smack dab through the middle of the Burghan oil field where a major portion of Kuwait's 980 burning oil wells were blazing away under a great black cloud that shut out the sun and rained a constant drizzle of tiny black tar balls. But that's another story.

When the cease-fire in that war was called after a hundred hours there were supplies in place for a hundred days, which for reasons only Uncle Sam knows could not be repatriated. That's a lot of good stuff; food, clothing, ammo, medical supplies etc. for an army of half a million. And then as luck would have it, our old enemy the USSR disappeared from the map and out of the rubble popped our new best friend Russia needing and pleading for all kinds of help. So, there we were, one of twelve 5 man teams, scattered throughout the former USSR arriving in a C141 (some of the lads in the even larger C5 Galaxy) loaded with leftover goodies from Gulf War I... Everything, except the ammo.

'VOLUNTEERS' OFF LOADING LEFTOVER GOODIES FROM GULF WAR ONE.

The lead agency at least for our group was OSIA, the On Site Inspection Agency, set up specifically to monitor and verify compliance with the Nuclear Disarmament Treaty extant at the time.

This allowed each side to make thirteen surprise visits annually to the other's Nuclear launch facilities. The visits followed an elaborate protocol designed to ensure surprise without getting shot by mistake. The inspecting team announced their arrival date ahead of time but did not specify the location. The 'when' but not the 'where', or vice versa. As the date approached the visiting side would narrow the site options only revealing the actual location a matter of hours before their arrival. All this to preserve the element of surprise while minimizing the chance of being shot down as an intruder. It also served to make it more difficult for the home team to prepare for the visit or, of course, to cheat.

According to our team leader, a USAF missileer, the two sides had very different approaches to compliance. The Soviets blew up their missiles while we, in the USA, dismantled ours.

Apart from there being nothing better to do in Siberia, the short but spectacular fireworks display from blowing up a fully fueled ICBM, was something none of our guys wanted to miss before signing off that the world was a little bit safer place. The Soviets on the other hand preferred to go shopping than watch an Atlas or Titan 2 being laboriously dismantled. Consequently the Soviet inspectors were bused to the nearest KMart and brought back only to inspect the scrap pile before signing off on our contribution to a better world.

Our team consisted of a missileer Air Force Major and team leader, an Army NCO, a Navy submariner, known as 'squids' in Navy jargon, both Russian language specialists, a G12 from the State Department and A.N.Other (myself) an alleged Logistics Expert.

We assembled in D.C. for a couple of days of briefing, issue of extreme weather kit, and some special gear before flying on to Rhine-Main Frankfurt. Among the special gear were a loaded money belt each, which only a strip search would find, and a suitcase with electronics and an umbrella device with which to contact Uncle Sam privately via the synchronous Indian Ocean satellite.

THE 'SQUID' SETTING UP FOR THE DAILY CHAT WITH UNCLE SAM IN D.C.

We overnighted in Rhein-main before flying to Moscow for a few days of briefings interspersed with sightseeing before heading out to Ulan Ude, Irkutsk and Chita where we were greeted with a welcome not seen since the end of World War 2.

'VOLUNTEERS' IN CHITA.

Each got a couple of MREs as payment. Best meal in months according to their commanding officer.

Our 'volunteers' at least in Chita, were all young, short and stocky. I mean 5'6" or less, prompting our fearless team leader, Major Slifka, to muse "there must be a tank regiment nearby".

Apparently Soviet tanks had 5 man crews packed into a small cabin with no power assists, so relying on muscle power.

The shell casings for example were brass, hot, and heavy, and the recoil could cut a mispositioned gunner in half. By contrast our cardboard casings were consumed in the firing chamber, only the base plug being ejected on the recoil.

Our 'volunteers' were rewarded with a couple of MREs per man.

A package containing everything needed to sustain a soldier for a day, with exotic sounding labels like 'Chicken Kiev or Beef Bourguignon'. The indestructible MRE has only one caveat: DO NOT DROP WHEN FROZEN. Everything in it is dehydrated and supposedly imperishable.

Just heat and eat, or for the dessert add water and watch.

I was particularly impressed, for example, with the fruit salad dessert, a perfect replica of those yellow-green pot scrubbers. Water added, it miraculously metamorphoses in moments into a fruit salad. . .

In addition to the multi thousand calories intake there was included provision for waste disposal. Matches, and toilet paper, are included for postprandial waste disposal... but our hungry volunteers were trying to eat everything so a training session was required.

Needless to say the explanation by our fearless leader, Major Slifka, of which orifice the TP was intended for, was both graphic and hilarious.

Picnic in Siberia,
engine hood for a hot plate.

In the cockpit of a Russian Antonov
Interpreter bottom right.

ICE FISHING AT 40 BELOW, LAKE BAIKAL

Footnotes:

*MRE: Meals Ready to Eat. Or, Meals Refused by Ethiopians as irreverent soldiers call them.

*RPG: Rocket propelled grenade. Fired from an AK47 to extend range for those w/o an outfielder's throwing arm.

**CIS: Confederation of Independent States. The residual rump of the Soviet Union after the defection of the East European and Central Asian 'Soviet States' under Boris Yeltsin.

A Very Special Train Ride

Siberia January 1992

On a parallel track, a freight train with flatbed wagons and hauling the latest Soviet T82 tanks, drew up alongside us, and for several minutes we traveled side by side at the same speed.

My companion in the cabin, Vladimir, was the head of the Trans-Siberian railway and a KGB general to boot.

Around my neck hung my Canon A1 camera and my trigger finger was itching to shoot it – the tank, that is – with my camera. Pleadingly, I looked at Vladimir, then at my camera, and then at the tank . . . saying nothing, while he glared at me with a bewildered look that seemed to say "a few weeks ago, I could have had you shot, but now I'm not sure." Then suddenly, he shrugged his shoulders and turned his back on me.

Soviet T82 tank with reactive armor

a four-pack on the turret

I fired off a few photos, dropped the camera back to its place around my neck, and re-engaged him in conversation about the history of the region or something equally innocuous. Having just returned from Gulf War One, I was particularly intrigued by what appeared to be reactive armor, something I don't recall seeing on our M1A1 Abrams tanks in Kuwait just a few months prior.

It was near the end of our mission to deliver leftover goodies from Gulf War One to the beleaguered remnants of the former Soviet Union, and Vladimir had arranged the special weekend trip, in an exclusive one-carriage train to Lake Baikal, the world's largest freshwater lake (holding more water than all the Great Lakes combined).

Our railcar was a VIP carriage with a bunch of sleeper cabins at one end; a kitchen in the middle, managed by a fearsome babushka (who could well have had daggers built into the toecaps of her all-weather, all-terrain boots); and a banquet room at the back. The banquet room still had all the paraphernalia of the Soviet era: red velvet curtains and seat covers, together with a large portrait of Lenin surveilling us all.

We had departed from the main terminal in Irkutsk and traveled for some time down the main East West line, which had multiple parallel tracks, before turning north on a single-track spur line that ran along the eastern shoreline of Lake Baikal. With the supreme commander of the Trans-Siberian railroad on board, it soon became apparent that the line had been closed to all other users for our sole and uninterrupted enjoyment!

The lake lies in a rift valley where the earth's crust is splitting. A mile deep, ten miles wide and almost 400 miles long, it is bounded by steep rocky terrain so the track follows a sequence of bridges and tunnels, ridges and valleys. Consequently we stopped frequently to explore places of interest: a Buddhist monastery; a Czarist-era mansion; a bridge; a tunnel. We also made an excursion onto the lake for a bare chested photo; and took rides in the locomotive and on the cowcatcher... all with photos, and, of course, with vodka to warm up when back in the train.

A Czarist mansion

A Buddhist monastery

Vladimir (crouching), with his family (l), myself, a female interpreter, and local

POLITICAL ESCORT (R)

BRIDGES, AND BARE CHESTED BRAVADO AT 40 BELOW

Far above the dramatic experience of such unfamiliar and exotic places, was the enduring warmth engendered between former mortal adversaries when helping, rather than harming, each other . . . and above all, relating to each other as human beings.

In the few weeks that culminated in that short weekend train ride, I believe that four and a half Americans and a handful of assorted Russians melted more Cold War ice, (per capita at least) than decades of diplomatic confrontation.

Well, maybe the vodka helped.

MUTAL RESPECT FROM FORMER ENEMIES.

Epilogue

On the twelfth of June 1987, speaking in front of the Berlin Wall, President Ronald Reagan challenged the Soviet Premier with these words: "Mister Gorbachov, tear down this wall."

The Wall divided East from West Berlin, them from us.

It was the physical manifestation of the "Iron Curtain", first articulated by Winston Churchill in Fulton Missouri almost half a century before.

Within months the wall came tumbling down, and with it the Soviet Union, and the Cold War was over.

Liberated from the threat of Mutually Assured Destruction, (MAD), the world rejoiced believing that perhaps we were all finally on the same side.

But it was not to be.

Human nature does not change overnight, and without the wall to divide them from us, (and us from them), we were soon inundated with their utopian ideology, as they were flooded with our promises of prosperity.

Alas, whereas red and blue blend into purple, oil and water don't mix and the Cold War ideological divide just shifted from the international to the domestic, from external to internal.

The clash of cultures metastasized like cancer into the fertile soil of the western democracies while the citizens of the Soviet Empire, freed from the straitjacket of conformity, rushed headlong into anarchy and crony capitalism in pursuit of prosperity.

The ruling elites, clinging to power and privilege, resorted to their favorite instrument of control, fear.

The ludicrous musical ditty 'Duck and Cover' that in the '50s conditioned school children to hide under their desks to survive a nuclear exchange, was replaced with the fear of climate catastrophe, global pandemics... and a future without hope.

Face masks, school closures and lockdowns replaced 'Duck and Cover'... but without even the jolly musical refrain.

So the process continues, Mother Nature will continue to afflict us with floods, famines, earthquakes, volcanoes and typhoons, and mankind will contribute war, civil or otherwise, to keep the disaster relief folks in business, for some blend of principle and profit.

Mother Nature, like an undisciplined child, will continue to wreak havoc in an attempt to get our attention.

Eventually, perhaps, the superior intelligence of Homo Sapiens will prevail over his inferior animal instinct to subdue others by force, and by so doing unleash the awesome power of human ingenuity to tame Mother Nature and harness her for the Greater Good.

But that's for another generation.

THE END.

for now…

Glossary

(*Words Mike's friends have heard him use over breakfast or cocktails, without ever really understanding what the heck he was talking about. - LMH*)

Askari	*The Apocalypse Now Gang* Soldier, or policeman, in both Swahili and Arabic
bedou	*Ali Subehi and Darfur Sudan* Collective term for several Bedouin (which is the singularform of the noun)
birr	*Tale of Two Nuns* Ethiopian currency
burp guns	*Saipan* The PPSh-41, a cheaper, simpler Soviet submachine gun, that the Russians nicknamed "daddy" but which was also called a "burp gun" due to its high fire-rate
Challenger Deep	*The Marianas* The deepest known point of the Earth's seabed, Challenger Deep was discovered by the survey ship HMS Challenger, part of the British Royal Navy, on its 1872-76 expedition to plumb the oceans.
Chokidar	*Tale of Two Nuns* A Rest House attendant
click	*Tale of Two Nuns* slang for 'kilometer'
Corniche	*The Colonel* The Bayfront road, Abu Dhabi

dhow	*Eat the Indian Mangoes* the generic name of a number of traditional sailing vessels with one or more masts, used in the Indian Ocean and Red Sea region.
dik dik	*The Lion King* A small, extremely shy antelope, seldom seen in the open, likes to hide in the bushes.
djini / djinn	*Reel in the Monkey, & Apocalypse Now Gang* a genie, or mischievous spirit
Eastern Carolinas	*The Marianas, Chasing Typhoons* Pacific atoll Islands now called the Federated States of Micronesia
fatoor	*Breakfast and the broken NavAid* from the Arabic root fatah – means "the opening," hence the "opening" meal of the day. Ironically, breakfast – "breaking the fast" – means the same thing
Enver Hoxha	*the Head Rebel* The hardline Stalinist dictator of Albania
Ghaffir	*Reel in the Monkey, Fish Out of water* A guard, or gate keeper. Typically an elderly man, well known and respected in the community, who carries a good stick and lies most of the time inside the gate on a wood framed hammock called a Rakuba. The stick, ostensibly to repel intruders, is often used to enhance his own mobility, especially as he gets older.

Term	Source / Definition
Haraami	*Reel in the Monkey* Means "Sinner" in Arabic, but is commonly used to describe thieves.
HRH Her Royal Highness	*Apocalypse Now Gang* For those unfamiliar with Royalty, the Monarch is addressed as 'Majesty' and members of the immediate family as 'Royal Highness'.
FLOSY	*Piper's Pies* Front for the Liberation of Occupied South Yemen
halal	*No Flies on My Man* Islamic ritual for preparing meat, the animal is bled to death by cutting the jugular
Hawajjah	*I am not Competent* A foreigner, especially a European, in colloquial Sudanese Arabic. Darfur, Sudan. Literally "one who travels".
Herc	*Lockheed C130 'Hercules' Cargo lifter* Heavy duty multipurpose military transport Aircraft.
Hose bib	*The Colonel* water hose.
Indian Wampas	*Tale of two Nuns* Native American decorative sea shells used as currency, similarly cowrie shells from the coast were once used as currency in the East African interior

JATO	*The Marianas* Jet Assisted Take Off: typically 3,000 pound (LOUD) thrust rockets that burn for a minute or so to bring a plane up to flying speed before being jettisoned
Jan-Jaweed	*The Apocalypse Now Gang* Gangs of Bedouin nomad raiders indigenous to Darfur Sudan.
Kali	*Fat Lion in a tree* Swahili for angry or bad-tempered.
L.R.A.	aka Lord's Resistance Army: gangs of armed poachers, and renegade army units from both Sudan and Uganda
Lebanese insurance	During the Lebanese Civil War, 1975 and onward, when assassins were available at $50 a pop, it was not uncommon for prominent individuals to set aside appropriate funds to ensure that should they have an unfortunate accident, those parties that might benefit therefrom would also have unfortunate accidents.. hence the term 'Lebanese Insurance,' a scaled-down localized version of the prevailing Mutually Assured Destruction (MAD) deterrence principle that happily preserved humanity throughout the Cold War... not to mention a few Lebanese businessmen!
Mamnooa	Forbidden, in Arabic, both Secular and Religious

mosey	*A one-armed Ethiopia*n To surreptitiously sidle over
murham	*Telephone Call from Mto wa Mbu* Red soil. (High iron content) used for red clay Tennis Courts and ubiquitous unpaved road surfaces in Africa. Becomes mud in the wet season, dust in the dry season.
m'zungu	*Fat lion* Swahili for a European or white guy born and raised in-country, fluent in the local language, and considered 'one of them', a local.
Nessun Dorma	*From Tigray to Typhoons* One of the best-known tenor arias in all of opera; from the final act of Giacomo Puccini's Turandot
NLF	*Aden, Pipers Pies* National Liberation Front
Qat. Aden	*The Party went off with a bang* A popular narcotic chew in Yemen and Somalia. The leaves of the plant contain cathinone, an alkaloid. A stimulant similar to cocaine. An 'upper'. Chewed not swallowed, it induces euphoria, erratic behavior, bug eyes, fast talk, a hamster like wad in the cheek and black teeth
Rakuba	*Reel in the Monkey* A wood-framed hammock for sleeping

Rorke's Drift	*A riot for Belafonte* A Battle in South Africa during the Anglo-Zulu War of 1879. 140 British Regulars and a handful of irregulars held off (just) an assault over two days by 3 to 4 thousand Zulus at the Swedish mission and former trading post of the Irishman James Rorke on the Mzinyathi river that divided the British colony of Natal from the Zulu kingdom. A drift is a ford or crossing. The battle is notable for the highest number of Victoria Crosses, (11) (the highest award for Gallantry in the British Military). Deemed a British victory because they held the fort and the Zulus retired with the loss of over 500 dead to 17 (and over 100 wounded) on the British side, the Zulu chief saluting the courage of his adversaries...so it was really a one sided draw . . . unlike General George Custer at the Little Big Horn three years earlier . . .
Sahil	*Darfur Sudar chapter* the semi-arid belt between the Sahara desert and the rain-fed regions to the south.
SAS	Special Air Service. Original British Special Forces regiment.

seamount	*Chasing Typhoons* A large submarine landform rising from the ocean floor that didn't quite reach the surface or sank below it from erosion or sea level rise. Volcanic in origin, when in the photosphere (sunlight), seamounts form the foundation for coral reefs which can adapt to sea level variations in response to global glaciation cycles. The result is a chain of Islands around a central lagoon, the old volcano's caldera, to create one of nature's most exotic and beautiful environs.
shifta	*The Lion King* Somali wildlife poachers. Bandits
souk	the market, in Arabic
Subehi	*Harry was my Boss* The South Yemen desert area between the coast and the highlands to the North
tabah	*A Riot for Belafonte* A large metal tray on which communal food is served.

tribe	The modern word "community" is not an adequate substitution because tribes are families, patriarchal and hierarchical. Each has its own boss, the sheik. Alliances with other tribes are typically through arranged marriages, the tribes do not merge. Abdul Aziz ibn Saud, united the tribes this way in the 1920's to create Saudi Arabia, which now has an estimated 5,000 Princes.... Princesses don't count..!
tukul	*The Lion King* round, palm thatched mud-and-wattle huts.
Umm	*Umm Dada* In Arabic, Umm means mother and Abu means father; each is used to denote a superlative. Saddam Hussein liked big battles (so long as others, not he, were getting killed) which he described as Umm Harb: Mother of War. When I was distributing Uncle Sam's famine relief (sorghum), in Darfur, my local moniker was Abu Aish . . . Father of food/ bread!
UNHCR	*Tiptoeing through the minefield* UN High Commission for Refugees
wadi	*Aden, & Darfur* a valley, dry river bed

wally	*Ali Subehi* a holy man, a saint or hermit; some living and others entombed, but still requiring an offering or blessing to ensure safe passage through their domain without a full complement (Kuwait 1991) not full; space available
WMD	weapons of mass destruction
Zulus	*A Riot for Belafonte* A powerful warrior tribe in South Africa

About the Author

Michael was born in New Zealand where his mother had returned home from Nigeria when, in the darkest hours of WW2, her British husband, a colonial administrator, was called up to defend the Empire. One Sunday morning en route to New Zealand from Cape Town the half dozen passengers were summoned urgently to the bridge by the ship's siren where the Captain announced that Pearl Harbor had been attacked.

Handing a glass of champagne to each of the bewildered passengers, he raised his glass with the toast "Thank God! Now the Yanks are in". A few months later with the 'Yanks' involved and the tide turning, his father, no longer needed to defend the Empire, was returned to his desk to administer it.

Tempting fate, Michael, with his mother set out from Auckland dodging the rampaging Japanese in the pacific and surviving three U boat wolf pack attacks in the North Atlantic that sank a third of their convoy, to arrive five months later in Liverpool, England, where at the age of thirteen months, he was introduced to his father.

Growing up in Nigeria before being farmed out to boarding schools in the UK, he eventually managed to graduate from Oxford University with a degree in Engineering Science. While at Oxford he joined the Air Squadron, an RAF reserve unit where he learned to fly (combat style) at the Queen's expense and signed up for a career in the RAF. Too tall to fly fighters he declined the offer to fly Bombers or Transports preferring to shoot at things rather than be shot at... and joined BP instead who sent him out to Aden in the Middle East where he was both shot at and bombed unsuccessfully.

Now retired he lives on Sanibel Island Florida keeping a close eye on his grandkids Sage and Mac, and their parents, to whom this book is dedicated... and playing golf with some of the best once used brains on the planet.

Cheers!

Printed in the USA
CPSIA information can be obtained
at www.ICGtesting.com
CBHW050120041024
15319CB00045B/750